Core 4

for Edexcel

CAMBRIDGE
UNIVERSITY PRESS

The School Mathematics Project

SMP AS/A2 Mathematics writing team Spencer Instone, John Ling, Paul Scruton, Susan Shilton, Heather West

SMP design and administration Melanie Bull, Pam Keetch, Nicky Lake, Cathy Syred, Ann White

The authors thank Sue Glover for the technical advice she gave when this AS/A2 project began and for her detailed editorial contribution to this book. The authors are also very grateful to those teachers who commented in detail on draft chapters.

CAMBRIDGE UNIVERSITY PRESS
Cambridge, New York, Melbourne, Madrid, Cape Town, Singapore, São Paulo

Cambridge University Press
The Edinburgh Building, Cambridge CB2 2RU, UK

www.cambridge.org
Information on this title: www.cambridge.org/9780521605380

© The School Mathematics Project 2005

First published 2005

Printed in the United Kingdom at the University Press, Cambridge

A catalogue record for this publication is available from the British Library

ISBN-13   978-0-521-60538-0 paperback
ISBN-10   0-521-60538-5 paperback

Typesetting and technical illustrations by The School Mathematics Project

The authors and publisher are grateful to London Qualifications Limited for permission to reproduce questions from past Edexcel examination papers. Individual questions are marked Edexcel. London Qualifications Limited accepts no responsibility whatsoever for the accuracy or method of working in the answers given.

Using this book

Each chapter begins with a **summary** of what the student is expected to learn.

The chapter then has sections lettered A, B, C, ... (see the contents overleaf). In most cases a section consists of development material, worked examples and an exercise.

The **development material** interweaves explanation with questions that involve the student in making sense of ideas and techniques. Development questions are labelled according to their section letter (A1, A2, ..., B1, B2, ...) and answers to them are provided.

D Some development questions are particularly suitable for discussion – either by the whole class or by smaller groups – because they have the potential to bring out a key issue or clarify a technique. Such **discussion questions** are marked with a bar, as here.

K **Key points** established in the development material are marked with a bar as here, so the student may readily refer to them during later work or revision. Each chapter's key points are also gathered together in a panel after the last lettered section.

The **worked examples** have been chosen to clarify ideas and techniques, and as models for students to follow in setting out their own work. Guidance for the student is in italic.

The **exercise** at the end of each lettered section is designed to consolidate the skills and understanding acquired earlier in the section. Unlike those in the development material, questions in the exercise are denoted by a number only.

Starred questions are more demanding.

After the lettered sections and the key points panel there may be a set of **mixed questions**, combining ideas from several sections in the chapter; these may also involve topics from earlier chapters.

Every chapter ends with a selection of **questions for self-assessment** ('Test yourself').

Included in the mixed questions and 'Test yourself' are **past Edexcel exam questions**, to give the student an idea of the style and standard that may be expected, and to build confidence.

Contents

1 Partial fractions

In this chapter you will learn how to
- add two or more algebraic fractions using the lowest common denominator
- express an algebraic fraction as the sum of simpler (partial) fractions

A Adding and subtracting algebraic fractions 1

To add or subtract fractions we write each fraction with a common denominator and using the lowest common denominator often simplifies the calculation. The lowest common denominator for the sum $\frac{1}{2} + \frac{1}{3}$ is the product $2 \times 3 = 6$: this gives $\frac{1}{2} + \frac{1}{3} = \frac{3}{6} + \frac{2}{6} = \frac{5}{6}$

In Core 3 we saw that, to add or subtract algebraic fractions to produce a single algebraic fraction, we must also first write each fraction with a common denominator.

Sometimes the lowest common denominator is the product of the denominators.

Example 1

Express $\dfrac{2}{x+4} - \dfrac{1}{2x+3}$ as a single algebraic fraction.

Solution

The lowest common denominator for $\dfrac{2}{x+4} - \dfrac{1}{2x+3}$ *is the product* $(x+4)(2x+3)$.

So $\dfrac{2}{x+4} - \dfrac{1}{2x+3} \equiv \dfrac{2(2x+3)}{(x+4)(2x+3)} - \dfrac{(x+4)}{(x+4)(2x+3)}$

$$\equiv \dfrac{2(2x+3)-(x+4)}{(x+4)(2x+3)}$$

$$\equiv \dfrac{4x+6-x-4}{(x+4)(2x+3)}$$

$$\equiv \dfrac{3x+2}{(x+4)(2x+3)}$$

Example 2

Express $\dfrac{1}{2x} + \dfrac{5}{3(2x-1)}$ as a single algebraic fraction.

Solution

The lowest common denominator for $\dfrac{1}{2x} + \dfrac{5}{3(2x-1)}$ *is* $2x \times 3(2x-1)$ *which is* $6x(2x-1)$.

So $\dfrac{1}{2x} + \dfrac{5}{3(2x-1)} \equiv \dfrac{3(2x-1)+10x}{6x(2x-1)}$

$$\equiv \dfrac{16x-3}{6x(2x-1)}$$

Example 3

Express $\dfrac{1}{x-1} - \dfrac{4}{x+2} + \dfrac{3}{x+3}$ as a single algebraic fraction.

Solution

The lowest common denominator for $\dfrac{1}{x-1} - \dfrac{4}{x+2} + \dfrac{3}{x+3}$ *is the product* $(x-1)(x+2)(x+3)$.

$$\frac{1}{x-1} - \frac{4}{x+2} + \frac{3}{x+3} \equiv \frac{(x+2)(x+3) - 4(x-1)(x+3) + 3(x-1)(x+2)}{(x-1)(x+2)(x+3)}$$

$$\equiv \frac{x^2 + 5x + 6 - 4(x^2 + 2x - 3) + 3(x^2 + x - 2)}{(x-1)(x+2)(x+3)}$$

$$\equiv \frac{x^2 + 5x + 6 - 4x^2 - 8x + 12 + 3x^2 + 3x - 6}{(x-1)(x+2)(x+3)}$$

$$\equiv \frac{12}{(x-1)(x+2)(x+3)}$$

Exercise A (answers p 152)

1 Express each of these as a single fraction in its simplest form.

(a) $\dfrac{1}{x-1} + \dfrac{2}{x+3}$ (b) $\dfrac{1}{x+2} - \dfrac{1}{x+5}$ (c) $\dfrac{3}{1+x} - \dfrac{1}{1-2x}$

(d) $\dfrac{1}{x+5} + \dfrac{1}{x+1} - \dfrac{2}{x+3}$ (e) $\dfrac{3}{5+x} - \dfrac{5}{2+x} + \dfrac{2}{x-4}$ (f) $\dfrac{3}{x-1} + \dfrac{1}{3-x} - \dfrac{1}{x+2}$

2 Express each of these as a single fraction in its simplest form.

(a) $\dfrac{3}{2x} + \dfrac{3}{x+1}$ (b) $\dfrac{2}{5} - \dfrac{1}{3(1-x)}$ (c) $\dfrac{1}{2(x+1)} + \dfrac{1}{3(x-2)}$

3 Express $5 - \dfrac{1}{2x-3}$ as a single fraction in its simplest form.

4 A function is defined by

$$\mathrm{f}(n) = \frac{1}{n-1} - \frac{2}{2n-1} \quad \text{where } n \text{ is an integer and } n \neq 1.$$

(a) Evaluate each of these in its simplest fractional form.

(i) $\mathrm{f}(2)$ (ii) $\mathrm{f}(4)$ (iii) $\mathrm{f}(10)$

(b) Prove that $\mathrm{f}(n)$ in its simplest form will always have a numerator of 1.

5 (a) Prove that $\dfrac{A}{x+1} + \dfrac{B}{x-1} \equiv \dfrac{(A+B)x + (B-A)}{x^2 - 1}$ for any constant values A and B.

(b) Hence write $\dfrac{5x+1}{x^2 - 1}$ in the form $\dfrac{A}{x+1} + \dfrac{B}{x-1}$.

B Adding and subtracting algebraic fractions 2

The product of the denominators for the sum $\frac{1}{4} + \frac{1}{6}$ is $4 \times 6 = 24$.
However this product does not give the **lowest** common denominator, which is 12.

Similarly, for two or more algebraic fractions the product of the denominators is not always the lowest common denominator.

Example 4

Express $\dfrac{1}{4x} + \dfrac{1}{6x^2}$ as a single algebraic fraction.

Solution

The product of $4x$ and $6x^2$ is $24x^3$ but a simpler common denominator is $12x^2$.

$$\frac{1}{4x} + \frac{1}{6x^2} \equiv \frac{3x + 2}{12x^2}$$

Example 5

Express $\dfrac{3}{2(x+1)} - \dfrac{x}{(x+1)^2}$ as a single algebraic fraction.

Solution

The lowest common denominator is $2(x + 1)^2$.

$$\frac{3}{2(x+1)} - \frac{x}{(x+1)^2} \equiv \frac{3(x+1) - 2x}{2(x+1)^2}$$

$$\equiv \frac{x+3}{2(x+1)^2}$$

Example 6

Express $\dfrac{2}{x+2} + \dfrac{1}{x-1} - \dfrac{3x}{(x-1)^2}$ as a single algebraic fraction.

Solution

The lowest common denominator is $(x + 2)(x - 1)^2$.

$$\frac{2}{x+2} + \frac{1}{x-1} - \frac{3x}{(x-1)^2} \equiv \frac{2(x-1)^2 + (x+2)(x-1) - 3x(x+2)}{(x+2)(x-1)^2}$$

$$\equiv \frac{2x^2 - 4x + 2 + x^2 + x - 2 - 3x^2 - 6x}{(x+2)(x-1)^2}$$

$$\equiv \frac{-9x}{(x+2)(x-1)^2}$$

Exercise B (answers p 152)

1 Express each of these as a single fraction in its simplest form.

(a) $\dfrac{1}{2x} - \dfrac{1}{4x}$ (b) $\dfrac{6}{x} - \dfrac{1}{x^2}$ (c) $\dfrac{5}{3x} + \dfrac{1}{x^2}$ (d) $\dfrac{1}{3x} - \dfrac{3}{2x^2}$

2 Express each of these as a single fraction in its simplest form.

(a) $\dfrac{3}{2(x-1)} + \dfrac{1}{4(x-1)}$ (b) $\dfrac{1}{10(x+2)} - \dfrac{2}{15(x-2)}$ (c) $\dfrac{1}{6x(x-1)} + \dfrac{1}{4(x-1)}$

3 Express each of these as a single fraction in its simplest form.

(a) $\dfrac{3}{x-1} + \dfrac{1}{(x-1)^2}$ (b) $\dfrac{1}{x+2} - \dfrac{2}{(x+2)^2}$ (c) $\dfrac{1}{2(x-1)} + \dfrac{2}{3(x-1)^2}$

(d) $\dfrac{1}{2(2x+1)} - \dfrac{x}{5(2x+1)^2}$ (e) $\dfrac{1}{3x+2} - \dfrac{2x+1}{(3x+2)^2}$ (f) $\dfrac{2}{x-3} + \dfrac{x+1}{(3-x)^2}$

4 (a) Show that $\dfrac{1}{x+2} - \dfrac{2}{x+4} + \dfrac{x}{(x+4)^2} \equiv \dfrac{-2x}{(x+2)(x+4)^2}$.

(b) Hence show that, when x is greater than 0, the value of

$\dfrac{1}{x+2} - \dfrac{2}{x+4} + \dfrac{x}{(x+4)^2}$ is negative.

5 Express each of these as a single fraction in its simplest form.

(a) $\dfrac{2}{2-x} + \dfrac{2}{x+1} - \dfrac{1}{(x+1)^2}$ (b) $\dfrac{4}{7(2x+1)} - \dfrac{2}{7(x-3)} + \dfrac{1}{(x-3)^2}$

(c) $\dfrac{1}{2(x+1)} - \dfrac{1}{3(x+2)} + \dfrac{1}{(x+2)^2}$ (d) $\dfrac{1}{x-1} + \dfrac{4}{(x-1)^2} + \dfrac{3}{(x-1)^3}$

(e) $\dfrac{1}{x+1} - \dfrac{2}{x+3} + \dfrac{x+2}{(x+3)^2}$ (f) $\dfrac{4}{3(1-x)} + \dfrac{3}{2x+1} + \dfrac{7-x}{(2x+1)^2}$

6 (a) Prove that $\dfrac{A}{x+2} + \dfrac{B}{(x+2)^2} \equiv \dfrac{Ax+(2A+B)}{(x+2)^2}$ for any constant values A and B.

(b) Hence write $\dfrac{2x+3}{(x+2)^2}$ in the form $\dfrac{A}{x+2} + \dfrac{B}{(x+2)^2}$.

C Partial fractions (answers p 153)

We can add any pair of algebraic fractions to produce a single algebraic fraction.
It is useful to be able to reverse this process, and later on in this course you will
do this to produce series expansions and solve some integration problems.
We shall first concentrate on the basic technique.

For example, the algebraic fraction $\dfrac{4}{(x+1)(x+3)}$ has a denominator with two linear factors.

Can we find a pair of fractions of the form $\dfrac{A}{x+1}$ and $\dfrac{B}{x+3}$ that add to give $\dfrac{4}{(x+1)(x+3)}$?

C1 (a) Show that $\dfrac{A}{x+1}+\dfrac{B}{x+3}\equiv\dfrac{(A+B)x+(3A+B)}{(x+1)(x+3)}$ for any constant values A and B.

(b) Given that $\dfrac{4}{(x+1)(x+3)}\equiv\dfrac{A}{x+1}+\dfrac{B}{x+3}$ explain why A and B must satisfy the simultaneous equations $A+B=0$ and $3A+B=4$.

(c) Solve the simultaneous equations $A+B=0$ and $3A+B=4$.

(d) Hence show that $\dfrac{4}{(x+1)(x+3)}\equiv\dfrac{2}{x+1}-\dfrac{2}{x+3}$.

C2 (a) Show that $\dfrac{A}{x-1}+\dfrac{B}{2x+5}\equiv\dfrac{(2A+B)x+(5A-B)}{(x-1)(2x+5)}$ for any constant values A and B.

(b) Hence write $\dfrac{7x}{(x-1)(2x+5)}$ in the form $\dfrac{A}{x-1}+\dfrac{B}{2x+5}$.

Example 7

Write $\dfrac{13}{(5+x)(3-2x)}$ in the form $\dfrac{A}{5+x}+\dfrac{B}{3-2x}$.

Solution

$$\dfrac{13}{(5+x)(3-2x)}\equiv\dfrac{A}{5+x}+\dfrac{B}{3-2x}$$

$$\equiv\dfrac{A(3-2x)+B(5+x)}{(5+x)(3-2x)}$$

$$\equiv\dfrac{3A-2Ax+5B+Bx}{(5+x)(3-2x)}$$

$$\Rightarrow\quad\dfrac{13}{(5+x)(3-2x)}\equiv\dfrac{(B-2A)x+(3A+5B)}{(5+x)(3-2x)}$$

The denominators are identical, and hence the numerators are too.

So we have these simultaneous equations. $\quad B-2A=0$ and $3A+5B=13$

Rearrange the first equation. $\quad B=2A$

Substitute in the second equation. $\quad 13A=13\Rightarrow A=1$

Substitute back in the first equation. $\quad B-2=0\Rightarrow B=2$

$$\text{So }\dfrac{13}{(5+x)(3-2x)}\equiv\dfrac{1}{5+x}+\dfrac{2}{3-2x}$$

K Writing $\dfrac{13}{(5+x)(3-2x)}$ in the form $\dfrac{1}{5+x}+\dfrac{2}{3-2x}$ is expressing it as **partial fractions**.

C3 Express each of these as partial fractions.

(a) $\dfrac{5x}{(x+2)(2x-1)}$ (b) $\dfrac{5}{(x-2)(x+3)}$ (c) $\dfrac{x+1}{(x-2)(x-5)}$ (d) $\dfrac{3}{(3-2x)(3+2x)}$

We can use a slightly different method that does not involve solving simultaneous equations, but involves substituting particular values for x.

For example, consider the algebraic fraction $\dfrac{24}{(x+1)(x-7)}$.

Assume that $\dfrac{24}{(x+1)(x-7)} \equiv \dfrac{A}{x+1}+\dfrac{B}{x-7}$, where A and B are constants to be found.

Then we have $\dfrac{24}{(x+1)(x-7)} \equiv \dfrac{A(x-7)+B(x+1)}{(x+1)(x-7)}$.

Now the denominators are identical, and hence the numerators must be identical.

So it must be true that $A(x-7)+B(x+1)=24$ for all values of x.

In particular, it must be true for $x=7$ (chosen to eliminate A), which gives

$$A(0)+B(7+1)=24$$
$$\Rightarrow \qquad 8B=24$$
$$\Rightarrow \qquad B=3$$

It must also be true for $x=-1$ (chosen to eliminate B), which gives

$$A(-1-7)+B(0)=24$$
$$\Rightarrow \qquad -8A=24$$
$$\Rightarrow \qquad A=-3$$

So $\dfrac{24}{(x+1)(x-7)} \equiv \dfrac{-3}{x+1}+\dfrac{3}{x-7}$, which we usually write as $\dfrac{3}{x-7}-\dfrac{3}{x+1}$.

C4 Use the 'substitution' method above to express these as partial fractions.

(a) $\dfrac{10}{(x-3)(x+2)}$ (b) $\dfrac{2(x-1)}{(x-5)(x+3)}$

C5 Use any method to express each of these as partial fractions.
You will need to factorise the denominators first.

(a) $\dfrac{1}{x^2+5x+6}$ (b) $\dfrac{x}{x^2-9}$ (c) $\dfrac{3x+1}{x^2-1}$ (d) $\dfrac{2x+5}{x^2+x}$

The method of substituting particular values for x is particularly useful when the denominator of an algebraic fraction has three linear factors.

For example, consider the algebraic fraction $\dfrac{3x-4}{(2x-1)(x+2)(x-3)}$.

Assume that $\dfrac{3x-4}{(2x-1)(x+2)(x-3)} = \dfrac{A}{2x-1} + \dfrac{B}{x+2} + \dfrac{C}{x-3}$ (A, B and C are constants).

Then we have $\dfrac{3x-4}{(2x-1)(x+2)(x-3)} = \dfrac{A(x+2)(x-3)+B(x-3)(2x-1)+C(x+2)(2x-1)}{(2x-1)(x+2)(x-3)}$.

Now the denominators are identical, and hence the numerators must be identical.

So $A(x+2)(x-3) + B(x-3)(2x-1) + C(x+2)(2x-1) = 3x-4$ for all values of x.

In particular, it must be true for $x = 3$ (chosen to eliminate A and B), which gives

$$A(0) + B(0) + C(3+2)(2\times3-1) = 3\times3-4$$
$$\Rightarrow \qquad\qquad 25C = 5$$
$$\Rightarrow \qquad\qquad C = \tfrac{5}{25} = \tfrac{1}{5}$$

It must also be true for $x = \tfrac{1}{2}$ (chosen to eliminate B and C), which gives

$$A\left(\tfrac{1}{2}+2\right)\left(\tfrac{1}{2}-3\right) + B(0) + C(0) = 3\times\tfrac{1}{2}-4$$
$$\Rightarrow \qquad\qquad -\tfrac{25}{4}A = -\tfrac{5}{2}$$
$$\Rightarrow \qquad\qquad A = -\tfrac{5}{2} \times -\tfrac{4}{25} = \tfrac{2}{5}$$

It must also be true for $x = -2$ (chosen to eliminate A and C), which gives

$$A(0) + B(-2-3)(2\times-2-1) + C(0) = 3\times-2-4$$
$$\Rightarrow \qquad\qquad 25B = -10$$
$$\Rightarrow \qquad\qquad B = -\tfrac{10}{25} = -\tfrac{2}{5}$$

So $\dfrac{3x-5}{(2x-1)(x+2)(x-3)} \equiv \dfrac{\tfrac{2}{5}}{2x-1} + \dfrac{-\tfrac{2}{5}}{x+2} + \dfrac{\tfrac{1}{5}}{x-3}$,

which we usually write as $\dfrac{2}{5(2x-1)} - \dfrac{2}{5(x+2)} + \dfrac{1}{5(x-3)}$.

C6 Express each of these as partial fractions.

(a) $\dfrac{3x^2-12x+11}{(x-1)(x-2)(x-3)}$

(b) $\dfrac{32-17x}{x(1-x)(x+4)}$

*C7 Show that $\dfrac{Ax+B}{(x+C)(x+D)} \equiv \dfrac{AC-B}{(C-D)(x+C)} + \dfrac{B-AD}{(C-D)(x+D)}$.

K A proper algebraic fraction is one where the degree of the polynomial that is the numerator is less than the degree of the polynomial that is the denominator.

Any proper algebraic fraction with a denominator that is a product of distinct linear factors can be written as partial fractions as the sum of proper fractions whose denominators are the linear factors.

Exercise C (answers p 153)

1 Express each of these as partial fractions.

(a) $\dfrac{3}{x(1-x)}$

(b) $\dfrac{3(x+10)}{x(x+6)}$

(c) $\dfrac{2x+7}{(x+3)(x+4)}$

(d) $\dfrac{x+5}{(x-3)(x+1)}$

(e) $\dfrac{5}{(2x+1)(3x-1)}$

(f) $\dfrac{6x}{(2-x)(2x-1)}$

(g) $\dfrac{2}{(x+1)(x+5)}$

(h) $\dfrac{x}{(x-1)(3x+1)}$

(i) $\dfrac{3x-1}{(x+2)(5x-1)}$

2 Express each of these as partial fractions.

(a) $\dfrac{4}{x^2-4}$

(b) $\dfrac{x+1}{x^2+3x}$

(c) $\dfrac{2x-1}{3x^2+5x+2}$

(d) $\dfrac{2}{4x^2-9}$

3 Express each of these as partial fractions.

(a) $\dfrac{5x^2+7}{(x-1)(x+2)(x+3)}$

(b) $\dfrac{6}{x(x+1)(x-2)}$

(c) $\dfrac{2x}{(x+1)(2x+1)(3x+1)}$

(d) $\dfrac{4x-1}{(x-1)(x+1)(2x+1)}$

4 A function is given by

$$\mathrm{f}(x) = \frac{5}{(x-3)(x+2)}, \quad x \in \mathbb{R}, \ x \neq -2, \ x \neq 3$$

(a) Express $\mathrm{f}(x)$ as partial fractions.

(b) Hence show that $\mathrm{f}'(x) = \dfrac{1}{(x+2)^2} - \dfrac{1}{(x-3)^2}$

*5 A function f is defined as $\mathrm{f}(n) = \dfrac{3}{1\times 4} + \dfrac{3}{4\times 7} + \dfrac{3}{7\times 10} + \ldots + \dfrac{3}{(3n-2)(3n+1)}$, where n is a positive integer.

(a) Evaluate $\mathrm{f}(3)$ and $\mathrm{f}(4)$ as fractions in their simplest form.

(b) (i) Express $\dfrac{3}{(3n-2)(3n+1)}$ as partial fractions.

(ii) Hence show that $\mathrm{f}(n) = 1 - \dfrac{1}{3n+1}$ and so $\mathrm{f}(n) < 1$ for all n.

D Improper fractions

An improper algebraic fraction is one where the degree of the polynomial that is the numerator is greater than or equal to the degree of the polynomial that is the denominator.

An improper fraction can be simplified by first dividing so that no improper fractions remain and then expressing any proper fractions as partial fractions where possible.

For example, consider the algebraic fraction $\dfrac{2x^2 + 3x - 1}{x^2 - 1}$.

We can first divide to obtain $\dfrac{2x^2 + 3x - 1}{x^2 - 1} \equiv \dfrac{2(x^2 - 1) + 3x + 1}{x^2 - 1} \equiv 2 + \dfrac{3x + 1}{x^2 - 1}$.

We can now express $\dfrac{3x + 1}{x^2 - 1}$ as $\dfrac{3x + 1}{(x - 1)(x + 1)}$, and writing this as partial fractions

gives $\dfrac{3x + 1}{x^2 - 1} \equiv \dfrac{2}{x - 1} + \dfrac{1}{x + 1}$.

So finally we have $\dfrac{2x^2 + 3x - 1}{x^2 - 1} \equiv 2 + \dfrac{2}{x - 1} + \dfrac{1}{x + 1}$.

Exercise D (answers p 154)

1 Find the values of the constants A, B and C in each case.

(a) $\dfrac{3x^2 + 6x - 2}{x^2 + x} \equiv A + \dfrac{B}{x} + \dfrac{C}{x + 1}$

(b) $\dfrac{2x^2 - 3x + 5}{x(x - 1)} \equiv A + \dfrac{B}{x} + \dfrac{C}{x - 1}$

(c) $\dfrac{8x^2}{(x - 3)(x + 5)} \equiv A + \dfrac{B}{x - 3} + \dfrac{C}{x + 5}$

(d) $\dfrac{4x^2 - 7x + 1}{2x^2 - x} \equiv A + \dfrac{B}{x} + \dfrac{C}{2x - 1}$

2 Find the values of the constants A, B, C and D where
$$\frac{2x^3 + x^2 - 6x - 12}{x^2 - 4} \equiv Ax + B + \frac{C}{x + 2} + \frac{D}{x - 2}$$

3 A function is given by
$$f(x) = \frac{3x^3 + 1}{x^2 - 1}, \quad x \in \mathbb{R}, \quad x \neq \pm 1$$

(a) Express $f(x)$ in the form $Ax + \dfrac{B}{x + 1} + \dfrac{C}{x - 1}$ where A, B and C are constants.

(b) Hence, or otherwise, show that $f'(x) < 3$ for all values of x in the domain.

E Further partial fractions (answers p 154)

E1 Express each of these as a single fraction in its simplest form.

(a) $\dfrac{1}{x} + \dfrac{1}{x^2 + 1}$

(b) $\dfrac{1}{x} - \dfrac{x - 1}{x^2 + 1}$

E2 (a) Can you find constants A and B such that $\dfrac{1-2x}{x\left(x^2+1\right)} \equiv \dfrac{A}{x} + \dfrac{B}{x^2+1}$?

What happens? Can you explain this?

(b) Can you find constants A, B and C such that $\dfrac{1-2x}{x\left(x^2+1\right)} \equiv \dfrac{A}{x} + \dfrac{Bx+C}{x^2+1}$?

If the denominator of an algebraic fraction is a product of expressions where one of them is a quadratic that cannot be factorised then it may not be possible to express the fraction as partial fractions where each numerator is a constant. Algebraic fractions of this type are beyond the scope of this course.

E3 Express $\dfrac{1}{x+3} - \dfrac{1}{(x+3)^2}$ as a single fraction in its simplest form.

E4 Explain why you cannot find constants A and B such that $\dfrac{x}{(x-1)^2} \equiv \dfrac{A}{x-1} + \dfrac{B}{x-1}$.

E5 (a) Can you find constants A and B such that

(i) $\dfrac{x}{(x-1)^2} \equiv \dfrac{A}{x-1} + \dfrac{B}{(x-1)^2}$ (ii) $\dfrac{3x+7}{(x+2)^2} \equiv \dfrac{A}{x+2} + \dfrac{B}{(x+2)^2}$

(b) (i) Show that $\dfrac{Ax+B}{(x+C)^2} \equiv \dfrac{A}{x+C} + \dfrac{B-AC}{(x+C)^2}$ where A, B and C are any constants.

(ii) Hence express $\dfrac{5x+16}{(x+3)^2}$ in the form $\dfrac{P}{x+3} + \dfrac{Q}{(x+3)^2}$.

E6 Express $\dfrac{3}{2x-5} + \dfrac{2}{x+1} + \dfrac{1}{(x+1)^2}$ as a single fraction in its simplest form.

K Any proper algebraic fraction with a denominator that is a product of linear factors, some of which are repeated twice, can be written as partial fractions where numerators are constant values and denominators are the linear factors and the squares of the repeated factors.

An example should help to clarify this.

Consider the algebraic fraction $\dfrac{x-8}{(x+1)(x-2)^2}$.

The denominator is $(x+1)(x-2)^2$, so the denominators of the partial fractions will be $(x+1)$, $(x-2)$ and $(x-2)^2$.

So we have $\dfrac{x-8}{(x+1)(x-2)^2} \equiv \dfrac{A}{x+1} + \dfrac{B}{x-2} + \dfrac{C}{(x-2)^2}$, where A, B and C are constants to be found.

Hence $\dfrac{x-8}{(x+1)(x-2)^2} \equiv \dfrac{A(x-2)^2 + B(x+1)(x-2) + C(x+1)}{(x+1)(x-2)^2}$.

So $A(x-2)^2 + B(x+1)(x-2) + C(x+1) \equiv x-8$ for all values of x.

In particular, it must be true for $x = 2$ (chosen to eliminate A and B), which gives

$$A(0) + B(0) + C(2+1) = 2 - 8$$

$$\Rightarrow \qquad 3C = -6$$

$$\Rightarrow \qquad C = -2$$

It must also be true for $x = -1$ (chosen to eliminate B and C), which gives

$$A(-1-2)^2 + B(0) + C(0) = -1 - 8$$

$$\Rightarrow \qquad 9A = -9$$

$$\Rightarrow \qquad A = -1$$

We cannot choose a value for x to eliminate A and C but we can choose a value to give an equation in A, B and C that we can solve to find the value of B.

A suitable value is $x = 3$, which gives $A(3-2)^2 + B(3+1)(3-2) + C(3+1) = 3 - 8$

$$\Rightarrow \qquad A + 4B + 4C = -5$$

$$\Rightarrow \qquad -1 + 4B - 8 = -5$$

$$\Rightarrow \qquad 4B = 4$$

$$\Rightarrow \qquad B = 1$$

Hence $\dfrac{x-8}{(x+1)(x-2)^2} \equiv -\dfrac{1}{x+1} + \dfrac{1}{x-2} - \dfrac{2}{(x-2)^2}.$

Exercise E (answers p 155)

1 Express each of these as partial fractions.

(a) $\dfrac{x+4}{(x+2)^2}$

(b) $\dfrac{8}{(x+3)(x+1)^2}$

(c) $\dfrac{x^2-3}{(x-2)(x-1)^2}$

(d) $\dfrac{3x+13}{(x-1)(x+3)^2}$

(e) $\dfrac{x^2+2}{(x-4)(x+2)^2}$

(f) $\dfrac{7x+4}{(x+4)(x-2)^2}$

(g) $\dfrac{x^2+2}{(x-5)(2-x)^2}$

(h) $\dfrac{13-2x^2}{(2x+1)(x+3)^2}$

(i) $\dfrac{7-x^2}{(2x-1)(x-5)^2}$

(j) $\dfrac{3-2x}{(x+2)(3x-1)^2}$

(k) $\dfrac{4x+3}{x(4x-1)^2}$

(l) $\dfrac{2x^2+5}{x^2(2x+5)}$

2 A function is given by

$$f(x) = \dfrac{8+5x}{(x-2)(x+1)^2}, \quad x \in \mathbb{R}, \ x \neq -1, \ x \neq 2$$

(a) Express $f(x)$ as partial fractions.

(b) Hence show that $f'(x) = 2\left(\dfrac{1}{(x+1)^3} + \dfrac{1}{(x+1)^2} - \dfrac{1}{(x-2)^2}\right).$

*3 A function is given by

$$f(x) = \frac{(8x+1)(x-2)}{(x-3)(2x-1)^2}, \quad x > 3$$

Show that f is decreasing for all values of x in the domain.

Key points

- Any proper algebraic fraction with distinct or repeated linear factors in the denominator can be expressed as partial fractions where the numerators are constants. For example,

$$\frac{5x+1}{(x-1)(2x+1)(x-5)} \quad \text{can be expressed in the form} \quad \frac{A}{x-1} + \frac{B}{2x+1} + \frac{C}{x-5}$$

and $\dfrac{5x+1}{(x-1)(2x+1)^2}$ can be expressed in the form $\dfrac{A}{x-1} + \dfrac{B}{2x+1} + \dfrac{C}{(2x+1)^2}$. (pp 11, 15)

Test yourself (answers p 156)

1 Write $\dfrac{1}{4(x+1)} + \dfrac{3}{4(x-1)} + \dfrac{1}{2(x-1)^2}$ as a single algebraic fraction in its simplest form.

2 Express $\dfrac{7-x}{(x+3)(2x+1)}$ in the form $\dfrac{A}{x+3} + \dfrac{B}{2x+1}$.

3 Express $\dfrac{7x+11}{(1+5x)(3-x)}$ as partial fractions.

4 Express $\dfrac{18}{(x+2)(x-1)(x-4)}$ as partial fractions.

5 A function is given by

$$f(x) = \frac{3(x+1)}{(x+2)(x-1)}, \quad x \in \mathbb{R}, \; x \neq -2, \; x \neq 1$$

(a) Express f(x) as partial fractions.

(b) Hence, or otherwise, prove that f$'(x) < 0$ for all values of x in the domain. Edexcel

6 Find the values of the constants A, B and C so that

$$\frac{4x^2 - 5x - 13}{(x+1)(x-3)} \equiv A + \frac{B}{x+1} + \frac{C}{x-3}$$

7 Express $\dfrac{x-1}{(2x+3)^2}$ as partial fractions.

8 Express $\dfrac{3x-1}{(2-3x)(1-x)^2}$ as partial fractions.

2 Parametric equations

In this chapter you will learn how to
- work with curves defined by two parametric equations, including the circle and ellipse
- convert between parametric and cartesian equations

A Coordinates in terms of a third variable (answers p 156)

Computer animators make objects on the screen change their position over time. An object's position at any moment can be given using (x, y) coordinates (also known as 'cartesian coordinates'). To instruct the computer to produce a required movement, the x-coordinate and y-coordinate can be separately defined in terms of time.

A1 A computer animator uses these equations to define the movement of a dot on the screen (x and y are in centimetres; t is in seconds).

$$x = 3t$$
$$y = 6t - t^2$$

(a) Copy this table and use the equations to complete it.

(b) On squared paper, using axes labelled x and y, plot the motion of the dot. What might the dot represent?

t	0	1	2	3	4	5	6
x		3					
y		5					

A2 Here a dot has been made to move in a straight line.

(a) By reading off values of x at $t = 1$, $t = 2$ and so on, state an equation for x in terms of t.

(b) Similarly, express y as a function of t.

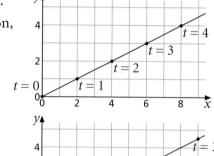

D **A3** Here, too, a dot moves in a straight line.

(a) How does the motion differ from that in question A2?

(b) Give an equation for x as a function of t.

(c) Give an equation for y as a function of t.

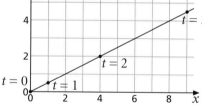

K Two equations that separately define the x- and y-coordinates of a graph in terms of a third variable are called **parametric equations**.

The third variable is called the **parameter**.

We have used t for the parameter here, but the parameter does not have to represent time, and a different letter could be used.

Many graph plotting calculators and programs will let you define a graph parametrically. The manual (often available online) should tell you how. You may be restricted to using t as the parameter.

You can also use a spreadsheet to plot a graph given by a pair of parametric equations, as here.

Column A contains values of t, increasing in steps of 0.2.
Column B gives the x-coordinate, defined by the function $x = t^3$.
Column C gives the y-coordinate, defined by the function $y = t^2$.

Enter −2

Enter =A2+0.2 and copy down as far as 2.

Enter =A2^3 and copy down.

Enter =A2^2 and copy down.

	A	B	C
	t	x	y
1			
2	−2	−8	4
3	−1.8	−5.832	3.24
4	−1.6	−4.096	2.56
5	−1.4	−2.744	1.96
6	−1.2	−1.728	1.44
7	−1	−1	1
8	−0.8	−0.512	0.64
9		−0.216	0.36

To plot the graph, first select the whole of columns B and C.
On the chart toolbar, select the 'scatter chart' button.
(In some versions of Excel, you have to make a 'First column contains …' selection; if so choose 'category (x)-axis labels' or 'x-values for xy-chart'.)

This method plots unjoined points for the graph. You may be able to obtain a scatter diagram with lines drawn between the points, but using a joined-up line chart option does not work with parametric equations on some spreadsheet programs.

A4 Obtain the graph given by $x = t^3$, $y = t^2$ $(-2 \le t \le 2)$ by plotting on squared paper or by using a graph plotter or spreadsheet. Call this graph G.

A5 Each of the following graphs is obtained by applying a transformation to graph G. In each case, plot the graph and describe the transformation.

(a) $x = t^3$, $y = t^2 + 2$ (b) $x = t^3 - 1$, $y = t^2$ (c) $x = 3t^3$, $y = t^2$

(d) $x = t^3$, $y = \frac{1}{2}t^2$ (e) $x = t^3$, $y = -t^2$ (f) $x = t^3 + 4$, $y = t^2 - 1$

An advantage of defining a graph parametrically is that transformations are simple to apply.

To apply a translation $\begin{bmatrix} a \\ b \end{bmatrix}$ add a to the function for x and b to the function for y.

To stretch in the x-direction, multiply the x-function by the required factor; similarly, to stretch in the y-direction, multiply the y-function.

To reflect in the y-axis multiply the x-function by -1; to reflect in the x-axis multiply the y-function by -1.

A6 State a pair of parametric equations for each of these.

(a) The graph $x = t^2 + t$, $y = 1 - t$ translated by $\begin{bmatrix} 3 \\ -1 \end{bmatrix}$

(b) The graph $x = t^2 + t$, $y = 1 - t$ stretched by a factor of 2 in the x-direction

Exercise A (answers p 156)

1 A curve is defined by $x = \dfrac{1}{t}$, $y = t^2$.

Find the coordinates of the points on the curve where $t = -3, -2, -1, 1, 2$ and 3.

2 A curve K is defined by $x = t$, $y = t^2$.

Give the parametric equations of a curve obtained by

(a) stretching K by factor 3 in the y-direction

(b) translating K by the vector $\begin{bmatrix} 2 \\ 1 \end{bmatrix}$

3 For each of the pairs of parametric equations below,

    (i) find the cartesian coordinates for $t = -2, -1, 0, 1, 2$, then plot these points and sketch the graph

    (ii) obtain the graph using a graph plotter or spreadsheet; if it differs from your sketch try to sort out why this has happened

(a) $x = t + 4$, $y = 1 - t^2$ (b) $x = 2 - t$, $y = t^3 - 2t$ (c) $x = t^3$, $y = t^2 - t$

4 A curve is defined by the parametric equations $x = 1 - t^2$, $y = 2t - 1$.

The point $(a, 9)$, where a is an unknown value, lies on this curve.

(a) Find the value of t at this point.

(b) Find the value of a.

5 A curve is defined by the parametric equations $x = t^3 + 1$, $y = \dfrac{t-1}{t+1}$.

The point $(-7, b)$ lies on this curve. Find the value of b.

6 A curve with the parametric equations $x = \dfrac{t}{2}$, $y = t^2 + t$ passes through

a point $(c, 12)$. Find two possible values for c.

*7 On the graphs below, the value of the parameter t is shown at each dot.
The parametric equations involve only low powers of t.
Write down possible parametric equations and test them on a graph plotter or spreadsheet.

(a)

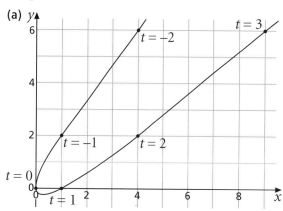

(b)

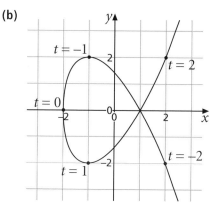

B Converting between parametric and cartesian equations (answers p 157)

B1 A graph is defined by these parametric equations.

$$x = 2t - 1$$
$$y = 4t$$

(a) Copy this table and complete it using the equations.

t	−2	−1.5	−1	−0.5	0	0.5	1	1.5	2
x									
y									

(b) Make t the subject of the second equation.

(c) Substitute this expression for t into the first equation, and show that this leads to

$$2x - y + 2 = 0$$

(d) Check that this equation is consistent with the last two lines of the table.

(e) What type of graph is this?

B2 Use the method of parts (b) and (c) in the previous question to convert each pair of parametric equations to an equation of the form $ax + by + c = 0$.

(a) $x = 3t + 4, \ y = 2t$ (b) $x = 5 - 2t, \ y = \frac{1}{3}t$

B3 Here again are the parametric equations from question A1:

$$x = 3t$$
$$y = 6t - t^2$$

(a) Make t the subject of the first equation.

(b) Show that substituting this expression for t into the second equation gives

$$y = 2x - \frac{x^2}{9}$$

(c) Check that the last two lines in your table from question A1 are consistent with this formula.

The method of questions B1–B3 can often be used to convert a pair of parametric equations into the equation that connects x and y directly (the cartesian equation). It involves first choosing the simpler equation and making t the subject of it.

B4 Obtain a single cartesian equation for each pair of parametric equations. Your equation does not have to be of the form $y = \ldots$

(a) $x = 5t, \ y = 4t - 3$ (b) $x = t^2, \ y = 2t$

(c) $x = 2t, \ y = t^3$ (d) $x = 3 - t, \ y = 3t$

(e) $x = 4t, \ y = 1 - t^2$ (f) $x = t^3, \ y = -t$

Example 1

Obtain a cartesian equation for the graph defined by $x = t^3$, $y = \sqrt{t}$ $(t \geq 0)$.

Solution

The second equation is the simpler.	$y = \sqrt{t}$
Square both sides.	$y^2 = t$
Substitute for t in the first equation.	$x = (y^2)^3 \implies x = y^6$

Example 2

Obtain a cartesian equation for the graph defined by $x = t + 5$, $y = t^3 - t$.

Solution

The first equation is the simpler.	$x = t + 5$
Make t the subject.	$t = x - 5$
Substitute for t in the second equation.	$y = (x - 5)^3 - (x - 5)$
Expand the brackets.	$y = x^3 - 15x^2 + 75x - 125 - x + 5$
Simplify.	$y = x^3 - 15x^2 + 74x - 120$

B5 Obtain a single cartesian equation for each pair of parametric equations.

(a) $x = \sqrt{t}$, $y = t^2$ (b) $x = t + 3$, $y = 3 - t^2$ (c) $x = t + 1$, $y = t^3 - 3t$

The following examples show a variety of ways of eliminating the parameter. In example 3, it is t^2, rather than t, that is first made the subject of one of the equations and eliminated. In example 4, the method is one that happens to work with the given equations. Example 5 again involves making t the subject, though the working is harder.

Example 3

Obtain a cartesian equation for the graph defined by $x = 3t^2 - 4$, $y = 8 - t^2$.

Solution

Take the y-equation and make t^2 the subject.	$t^2 = 8 - y$
Substitute $8 - y$ for t^2 in the x-equation.	$x = 3(8 - y) - 4$
Tidy the equation.	$x = 20 - 3y$

Example 4

Obtain a cartesian equation for the graph defined by $x = t + \dfrac{1}{t}$, $y = t - \dfrac{1}{t}$ $(t \neq 0)$.

Solution

Add the two parametric equations.	$x + y = 2t$
Subtract the second parametric equation from the first.	$x - y = \dfrac{2}{t}$
Multiply the 'sum' equation by the 'difference' equation.	$(x + y)(x - y) = 4$, or $x^2 - y^2 = 4$

Example 5

Obtain a cartesian equation for the graph defined by $x = \dfrac{t}{t+1}$, $y = \dfrac{t}{t-3}$ $(t \neq -1, t \neq 3)$.

Solution

Multiply both sides of the x-equation by t + 1.

$$tx + x = t$$

Bring all the t-terms together.

$$tx - t = -x$$

$$\Rightarrow \quad t(x - 1) = -x$$

$$\Rightarrow \quad t = \frac{-x}{x-1}$$

You will also need an expression for t − 3.

$$t - 3 = \frac{-x}{x-1} - 3$$

$$= \frac{-x}{x-1} - \frac{3(x-1)}{x-1}$$

$$= \frac{-x - 3(x-1)}{x-1}$$

$$= \frac{-4x + 3}{x-1}$$

Substitute the expressions for t and t − 3 into the y-equation.

$$y = \frac{-x}{x-1} \div \frac{-4x+3}{x-1} = \frac{-x}{x-1} \times \frac{x-1}{-4x+3} = \frac{-x}{-4x+3}$$

So the required cartesian equation is $y = \dfrac{-x}{-4x+3}$.

A statement like $t \neq -2$ or $t \geq 0$ after a pair of parametric equations is often warning you about a value or values of the parameter for which one or more of the equations is invalid (perhaps because division by zero or finding the square root of a negative number would be involved). You do not normally have to do anything about this.

Exercise B (answers p 157)

1 Obtain a single cartesian equation for each pair of parametric equations.

(a) $x = \frac{1}{4}t$, $y = 5t - 1$

(b) $x = 4t$, $y = \dfrac{4}{t}$ $(t \neq 0)$

(c) $x = t^2 + 4t$, $y = \frac{1}{3}t$

(d) $x = \dfrac{1}{t}$, $y = \frac{1}{2}t$ $(t \neq 0)$

(e) $x = 2 - t$, $y = t^2 + 4$

(f) $x = 2t - 1$, $y = 3 - 4t$

(g) $x = \sqrt{t}$, $y = t^2 + t$ $(t \geq 0)$

(h) $x = \dfrac{1}{t}$, $y = 4 - t$ $(t \neq 0)$

(i) $x = t^3 - t$, $y = t + 2$

(j) $x = 1 + t$, $y = 2t^2 - t^3$

(k) $x = \dfrac{t-1}{2}$, $y = (t+1)(t+2)$

(l) $x = 4t^3 + 3$, $y = 6 - t^3$

2 A curve is defined parametrically by $x = t^3 + \dfrac{3}{t}$, $y = t^3 - \dfrac{3}{t}$ $(t \neq 0)$.

By first expressing $x + y$ and $x - y$ in terms of t, show that $(x + y)(x - y)^3 = 432$.

3 Obtain a single cartesian equation for each pair of parametric functions.

(a) $x = \sqrt{t}$, $y = t(5 - t)$ $(t \geq 0)$

(b) $x = \dfrac{1}{t - 3}$, $y = t^2$ $(t \neq 3)$

(c) $x = \dfrac{1}{t - 2}$, $y = t^3$ $(t \neq 2)$

(d) $x = \dfrac{1}{t + 2}$, $y = \dfrac{1}{t - 1}$ $(t \neq -2, t \neq 1)$

(e) $x = 5 - \sqrt{t}$, $y = 4\sqrt{t} - 3$ $(t \geq 0)$ (f) $x = \dfrac{t}{2t - 1}$, $y = \dfrac{t}{t - 1}$ $(t \neq \frac{1}{2}, t \neq 1)$

4 A curve is defined by $x = \dfrac{1}{t}$, $y = \dfrac{1}{t(t - 1)}$. By first expressing $x + y$ and $\dfrac{y}{x}$ in terms of t, find the cartesian equation.

*5 Show that the graph defined by $x = \dfrac{1}{2t - 1}$, $y = \dfrac{t}{2t - 1}$ is a straight line.

C Solving problems (answers p 158)

Often in mathematics you can deal with a problem by forming an equation (or a pair of simultaneous equations) and then solving by the methods you used in Core 1.

Suppose you are told that a graph is defined by a pair of equations of the form $x = t + a$, $y = at$, where t is a parameter and a is a constant.
You are asked to find the value of a, given that the graph goes through $(6, 9)$.

As $(6, 9)$ is on the graph you can write these two equations:

$6 = t + a$
$9 = at$

C1 (a) Make t the subject of the first equation.

(b) Substitute your expression for t into the second equation, and show that what you obtain can be rearranged into $a^2 - 6a + 9 = 0$.

(c) Solve this equation to find the value of a.

C2 A curve defined by the parametric equations $x = pt^2$, $y = 3pt$, where t is a parameter and p is a constant, passes through the point $(-12, 18)$.
By forming and solving a pair of simultaneous equations, find the value of p.

C3 A curve defined by the parametric equations $x = t - b$, $y = 2b - t$ goes through the point $(-3, 5)$.

(a) Find the value of the constant b.

(b) Find the value of the parameter t at the point $(-3, 5)$.

C4 A curve defined by the parametric equations $x = 3t - q$, $y = t^2 + q$ goes through the point $(-13, 13)$. Find two possible values of the constant q.

Finding an unknown constant in a pair of parametric equations, given that the curve they define goes through a certain point, may be a simpler task if one coordinate of the point is zero. The following example illustrates this.

Example 6

A curve defined by the parametric equations $x = p(t^3 - 8)$, $y = pt$, where p is a constant, passes through the point $(0, -6)$. Find the value of p.

Solution

Write a pair of simultaneous equations.

$$0 = p(t^3 - 8) \qquad (1)$$
$$-6 = pt \qquad (2)$$

From equation (1), either $p = 0$ or $(t^3 - 8) = 0$.

But the first alternative cannot be true if equation (2) is true, so

$$(t^3 - 8) = 0$$
$$\Rightarrow \qquad t^3 = 8$$
$$\Rightarrow \qquad t = 2 \quad \textit{giving the value of the parameter at } (0, -6)$$

Substituting this value for t in equation (2),

$$-6 = 2p$$
$$\Rightarrow \qquad p = -3$$

A different problem is to find where a parametrically defined graph meets an axis, or a line parallel to an axis. This can be done by forming and solving a single equation as follows.

Example 7

Find the coordinates of the points where the curve $x = t^3 + 4$, $y = t^2 - t$ meets the line $y = 12$.

Solution

Form an equation in t that will be true at the required points then solve for t.

Substituting 12 for y in the y-equation,

$$12 = t^2 - t$$
$$\Rightarrow \qquad t^2 - t - 12 = 0$$
$$\Rightarrow \qquad (t - 4)(t + 3) = 0$$
$$\Rightarrow \qquad t = 4 \text{ or } -3$$

Substitute these values of t into the x-equation.

When $t = 4$, $x = 4^3 + 4 = 68$; when $t = -3$, $x = (-3)^3 + 4 = -23$.

So the points are $(68, 12)$ and $(-23, 12)$.

You can find a point where two graphs meet, when one of the graphs is defined by parametric equations, without converting them to a cartesian equation, as in the following example.

Example 8

A curve defined by the parametric equations $x = t^2$, $y = 2(t + 1)$ intersects the line $y = 2(x - 1)$ at two points. Find the coordinates of each point.

Solution

Form an equation in t that will be true at the two points then solve for t.

Substituting t^2 for x and $2(t + 1)$ for y in the equation $y = 2(x - 1)$,

$$2(t + 1) = 2(t^2 - 1)$$

$\Rightarrow \qquad t + 1 = t^2 - 1$

$\Rightarrow \qquad t^2 - t - 2 = 0$

$\Rightarrow \quad (t + 1)(t - 2) = 0$

$\Rightarrow \qquad\qquad t = -1 \text{ or } 2$

Use the parametric equations.

When $t = -1$, $x = (-1)^2 = 1$ and $y = 2(-1 + 1) = 0$, so $(1, 0)$ is a point of intersection.

When $t = 2$, $x = 2^2 = 4$ and $y = 2(2 + 1) = 6$, so $(4, 6)$ is the other point of intersection.

Exercise C (answers p 158)

1 A curve defined by the parametric equations $x = 4at$, $y = at^2$, where a is a constant, passes through the point $(8, 1)$. Find the value of a.

2 A curve defined by the parametric equations $x = t + m$, $y = \dfrac{t}{m}$ goes through the point $(9, 2)$. Find the value of the constant m.

3 A curve defined by the parametric equations $x = t + b$, $y = 3b - t$ goes through the point $(2, 10)$. Find the value of the constant b.

4 A curve defined by the parametric equations $x = t^2 - k$, $y = 5t + k$ goes through the point $(30, 20)$. Find two possible values of the constant k.

5 Use the fact that $x = 0$ for all points on the y-axis to find where the curve defined by $x = t^2 - 4$, $y = t^3 + t$ meets the y-axis.

6 Find the coordinates of the point(s) where each of the following curves meets the y-axis.

(a) $x = t - 2$, $y = 2t + 1$ (b) $x = 3 - t$, $y = t^2 - t$

(c) $x = t^2 + t - 2$, $y = 3t - 6$ (d) $x = 8 - t^3$, $y = \dfrac{t-1}{t+1}$

7 Find the coordinates of the points where the curves in question 6 meet the x-axis.

8 Find the coordinates of the point of intersection of the line $x = \dfrac{t}{2}$, $y = t + 1$ and the line $x + 2y = 5$.

9 Find the coordinates of the point where the curve $x = t + \dfrac{1}{t}$, $y = t - \dfrac{1}{t}$ intersects the line $y = x - 1$.

10 Show that the curve $x = \sqrt{t + 1}$, $y = \dfrac{t}{t^2 - 2}$ meets the line $y = 1$ at $(0, 1)$ and $(\sqrt{3}, 1)$.

11 The curve $x = t^2 + 1$, $y = t^3 - 1$ meets the line $x = 3$ at points A and B. Find the exact length of the line segment AB.

12 A curve defined by the parametric equations $x = m(1 - t^2)$, $y = m(3t - 5)$, where m is a constant, passes through the point $(-16, 0)$. Find the value of m.

13 A curve defined by the parametric equations $x = q(2t + 1)$, $y = q(t^3 + 27)$ passes through the point $(10, 0)$. Find the value of the constant q.

14 A curve is defined by the parametric equations $x = b(1 - 3t)$, $y = b(t^2 + t)$, where b is a constant. The curve passes through $(0, 4)$.

  (a) Find the value of b.

  (b) Find the coordinates of the points where the curve meets the x-axis.

15 Find the coordinates of the points where the line $x = t - 1$, $y = 2t - 3$ intersects the circle $x^2 + y^2 - 6x - 16 = 0$.

16 Find the exact coordinates of the points where the parabola $x = t$, $y = t^2 - 1$ meets the circle $x^2 + y^2 - 2y - 3 = 0$.

D Circle and ellipse (answers p 159)

You know from your work on trigonometry that $\cos\theta$ and $\sin\theta$ are defined as the x- and y-coordinates of a point rotating around a circle of radius 1 unit:

$x = \cos\theta$, $y = \sin\theta$

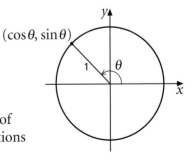

You can also regard these equations as defining x and y in terms of a third value θ. Seen this way, they are a pair of parametric equations for a circle, with θ as the parameter.

  D1 Put $x = \cos\theta$, $y = \sin\theta$ into a graph plotter and check that you get a circle with its centre at the origin.

  D2 What do you expect these pairs of parametric equations to give? Check with a graph plotter.

  (a) $x = 3\cos\theta$, $y = 3\sin\theta$ **(b)** $x = 0.5\cos\theta$, $y = 0.5\sin\theta$

In section A we applied a translation to a parametrically defined graph by adding a constant to the x-equation or the y-equation (or both).

D3 Write parametric equations for each of these, then check on a graph plotter.

    (a) A circle with unit radius, centre $(0, 2)$

    (b) A circle with unit radius, centre $(-3, 0)$

    (c) A circle with unit radius, centre $(1, -6)$

    (d) A circle with radius 2 units, centre $(5, 4)$

The curve $x = r\cos\theta$, $y = r\sin\theta$ is a circle with radius r, centre the origin.

The curve $x = r\cos\theta + p$, $y = r\sin\theta + q$ is a circle with radius r, centre (p, q).

D4 Sketch these circles, indicating the radius and the position of the centre in each case.

    (a) $x = \cos\theta - 3$, $y = \sin\theta + 2$ **(b)** $x = \cos\theta + 1$, $y = \sin\theta$

    (c) $x = 2\cos\theta - 5$, $y = 2\sin\theta + 5$ **(d)** $x = 0.6\cos\theta$, $y = 0.6\sin\theta - 3$

D5 Put each of these pairs of equations into a graph plotter. What transformation of the circle $x = \cos\theta$, $y = \sin\theta$ do you obtain in each case?

    (a) $x = \cos\theta$, $y = 2\sin\theta$ **(b)** $x = 4\cos\theta$, $y = \sin\theta$

    (c) $x = \cos\theta$, $y = 0.6\sin\theta$ **(d)** $x = 0.5\cos\theta$, $y = 1.2\sin\theta$

When you stretch a circle with unit radius, centre the origin, by factors of a in the x-direction and b in the y-direction you get an **ellipse** that cuts the x-axis at $(-a, 0)$ and $(a, 0)$ and cuts the y-axis at $(0, b)$ and $(0, -b)$.

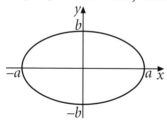

The curve $x = a\cos\theta$, $y = b\sin\theta$ is an ellipse, centre the origin. Its width is $2a$ units and its height is $2b$ units.

D6 Write a pair of parametric equations for each of these ellipses.

(a)

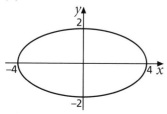

(b)

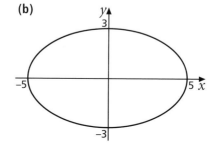

(c)

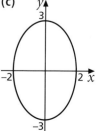

Like a circle, an ellipse can also be defined parametrically when its centre is not the origin.

D7 Write a pair of parametric equations for each of these ellipses.

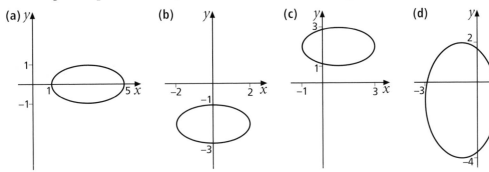

To convert a pair of parametric equations for an ellipse to a cartesian equation, first obtain expressions for $\cos\theta$ and $\sin\theta$ from the parametric equations, then use the identity $\cos^2\theta + \sin^2\theta = 1$, as in the next example.

Example 9

An ellipse is defined by the parametric equations $x = 4\cos\theta$, $y = 5\sin\theta$.
Find its cartesian equation.

Solution

From the x-equation, $\cos\theta = \dfrac{x}{4}$ From the y-equation, $\sin\theta = \dfrac{y}{5}$

Substituting these expressions in $\cos^2\theta + \sin^2\theta = 1$, $\left(\dfrac{x}{4}\right)^2 + \left(\dfrac{y}{5}\right)^2 = 1$

So the cartesian equation is $\dfrac{x^2}{16} + \dfrac{y^2}{25} = 1$.

D8 Use the method of example 9 to show that the general ellipse, defined by $x = a\cos\theta$, $y = b\sin\theta$, has the cartesian equation $\dfrac{x^2}{a^2} + \dfrac{y^2}{b^2} = 1$.

D9 An ellipse has the equation $9x^2 + 4y^2 = 36$.
Rewrite this in the form $\dfrac{x^2}{a^2} + \dfrac{y^2}{b^2} = 1$ and hence write down a pair of parametric equations for this ellipse.

D10 Find a pair of parametric equations for each of these ellipses.
    (a) $4x^2 + y^2 = 36$ **(b)** $x^2 + 9y^2 = 9$ **(c)** $9x^2 + 0.25y^2 = 2.25$

Example 10

A circle is defined by the parametric equations $x = 3\cos\theta + 2$, $y = 3\sin\theta - 1$.
Obtain its cartesian equation.

Solution

Make $\cos\theta$ and $\sin\theta$ the subject of the respective parametric equations.

$$\cos\theta = \frac{x-2}{3}, \quad \sin\theta = \frac{y+1}{3}$$

Substitute these expressions into $\cos^2\theta + \sin^2\theta = 1$.

$$\left(\frac{x-2}{3}\right)^2 + \left(\frac{y+1}{3}\right)^2 = 1$$

Hence the cartesian equation of the circle is $(x-2)^2 + (y+1)^2 = 3^2$.

This can be simplified to $x^2 + y^2 - 4x + 2y - 4 = 0$.

Notice that the above cartesian equation $(x-2)^2 + (y+1)^2 = 3^2$ is of the standard form $(x-a)^2 + (y-b)^2 = r^2$ for a circle with centre (a, b) and radius r (see Core 2), where $a = 2$, $b = -1$ and $r = 3$. These values are exactly what you would expect from the given parametric equations.

Exercise D (answers p 159)

1 A circle is defined by $x = 2\cos\theta$, $y = 2\sin\theta$, where θ is in radians.
Give the exact values of the cartesian coordinates of the point where $\theta = \dfrac{\pi}{3}$.

2 Write a pair of parametric equations for each ellipse produced as follows.

(a) From a circle, centre the origin and of unit radius, that has been stretched by a factor of 4 in the x-direction

(b) From a circle, centre the origin and of unit radius, that has been 'stretched' by a factor of 0.7 in the y-direction

3 An ellipse is defined by the parametric equations $x = 3\cos\theta$, $y = 4\sin\theta$.
Find the exact coordinates of the points where θ has values $0, \dfrac{\pi}{4}, \dfrac{\pi}{2}, \pi$ radians.

4 Find the cartesian equation of the ellipse defined by $x = \frac{1}{2}\cos\theta$, $y = \frac{1}{3}\sin\theta$.

5 The curves shown are ellipses. In each case, write

  (i) a pair of parametric equations   (ii) the cartesian equation

(a)

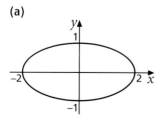

(b)

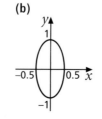

(c)

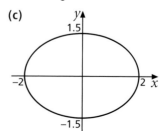

6 A circle is defined by the cartesian equation $(x-3)^2 + (y+2)^2 = 25$.

    (a) State its centre.   **(b)** State its radius.   **(c)** Define it parametrically.

7 For each of the following pairs of parametric equations,

    (i) fully describe the curve defined

    (ii) obtain the cartesian equation for the curve

    (a) $x = 2\cos\theta,\ y = 3\sin\theta$ **(b)** $x = \cos\theta,\ y = \sin\theta + 1$

    (c) $x = 4\cos\theta,\ y = 4\sin\theta$ **(d)** $x = \cos\theta + 1,\ y = 2\sin\theta$

    (e) $x = 2\cos\theta - 3,\ y = 2\sin\theta$ **(f)** $x = 2\cos\theta + 3,\ y = 4\sin\theta + 2$

    (g) $x = 5\cos\theta - 2,\ y = 5\sin\theta + 3$ **(h)** $x = 0.5\cos\theta - 3,\ y = 2\sin\theta + 1$

*8 Define the circle $x^2 + y^2 - 6x + 8y - 11 = 0$ parametrically.

E Using other trigonometric identities

In section D you used the identity $\cos^2\theta + \sin^2\theta = 1$ when manipulating parametric equations. The identities below may also be used.

Key points from Core 3

- $\sec\theta = \dfrac{1}{\cos\theta}$ $\operatorname{cosec}\theta = \dfrac{1}{\sin\theta}$ $\cot\theta = \dfrac{1}{\tan\theta}$

- $\sec^2\theta = 1 + \tan^2\theta$ $\operatorname{cosec}^2\theta = 1 + \cot^2\theta$

- $\sin 2\theta = 2\sin\theta\cos\theta$

- $\cos 2\theta = \cos^2\theta - \sin^2\theta = 2\cos^2\theta - 1 = 1 - 2\sin^2\theta$

- $\tan 2\theta = \dfrac{2\tan\theta}{1 - \tan^2\theta}$

Example 11

A curve is defined by the parametric equations $x = \frac{1}{2}\cos\theta,\ y = 4\sec\theta$.
Obtain a cartesian equation for this curve.

Solution

Make $\cos\theta$ and $\sec\theta$ the subject of the respective parametric equations.

    $\cos\theta = 2x,\ \sec\theta = \frac{1}{4}y$

Choose an identity that has $\cos$ and $\sec$ in it.

    $\sec\theta = \dfrac{1}{\cos\theta}$

Substitute the expressions for $\cos\theta$ and $\sec\theta$ that you have just obtained.

    $\frac{1}{4}y = \dfrac{1}{2x}$, which simplifies to the cartesian equation $y = \dfrac{2}{x}$

Example 12

Obtain a cartesian equation for the curve defined by the parametric equations
$x = \mathrm{cosec}\,\theta + 1, \; y = 3\cot\theta$.

Solution

Make $\mathrm{cosec}\,\theta$ *and* $\cot\theta$ *the subject of the respective parametric equations.*

$$\mathrm{cosec}\,\theta = x - 1, \; \cot\theta = \tfrac{1}{3}y$$

Choose an identity that has cosec *and* cot *in it.*

$$\mathrm{cosec}^2\,\theta = 1 + \cot^2\theta$$

Substitute the expressions for $\mathrm{cosec}\,\theta$ *and* $\cot\theta$ *that you have just obtained.*

$$(x - 1)^2 = 1 + (\tfrac{1}{3}y)^2$$

Simplify this cartesian equation.

$$x^2 - 2x + 1 = 1 + \tfrac{1}{9}y^2 \; \Rightarrow \; y^2 = 9x^2 - 18x$$

Example 13

Obtain a cartesian equation for the curve defined by the parametric equations
$x = \sin 2\theta + 1, \; y = \cos\theta \; (0 \le \theta \le \pi)$.

Solution

Use a double angle identity to rewrite the x-equation in terms of single θs.

$$x = 2\sin\theta\cos\theta + 1$$

You have a mixture of $\sin\theta$ *and* $\cos\theta$ *here but only* $\cos\theta$ *in the y-equation. It will help if you eliminate* $\sin\theta$. *Do this by first choosing an identity that contains* $\sin\theta$ *and* $\cos\theta$ *and writing* $\sin\theta$ *in terms of* $\cos\theta$.

$$\cos^2\theta + \sin^2\theta = 1$$

$$\Rightarrow \quad \sin^2\theta = 1 - \cos^2\theta$$

$$\Rightarrow \quad \sin\theta = \sqrt{1 - \cos^2\theta} \quad \textit{ignoring the negative square root because } 0 \le \theta \le \pi$$

Substitute this expression for $\sin\theta$ *in the rewritten x-equation.*

$$x = 2\left(\sqrt{1 - \cos^2\theta}\right)\cos\theta + 1$$

From the parametric equation for y, substitute y for $\cos\theta$ *to get the cartesian equation.*

$$x = 2\left(\sqrt{1 - y^2}\right)y + 1, \; \text{ or } \; x = 2y\sqrt{1 - y^2} + 1$$

Exercise E (answers p 160)

1 Obtain a single cartesian equation for each pair of parametric equations.

(a) $x = 2\,\mathrm{cosec}\,\theta, \; y = \sin\theta$ (b) $x = \tan\theta, \; y = \tfrac{1}{2}\cot\theta$

(c) $x = \sec^2\theta, \; y = \cos\theta$ (d) $x = \mathrm{cosec}\,\theta + 2, \; y = 4\sin\theta$

(e) $x = 5\cos\theta, \; y = \cos 2\theta$ (f) $x = \sin\theta, \; y = 3\cos 2\theta$

2 Obtain a single cartesian equation for each pair of parametric equations.

(a) $x = \cot\theta$, $y = 2\csc\theta$

(b) $x = \sin\theta$, $y = \sin 2\theta$ $\left(0 \le \theta \le \dfrac{\pi}{2}\right)$

(c) $x = 2\sec\theta$, $y = 3\tan\theta$

(d) $x = \tan\theta - 3$, $y = 3\sec\theta$

(e) $x = 4\cot\theta + 1$, $y = 2\csc\theta - 2$

(f) $x = \tan 2\theta$, $y = \tan\theta + 1$

(g) $x = \tan\theta$, $y = \cos\theta$

(h) $x = \tan\theta$, $y = 2\csc\theta$

3 A curve is defined by the parametric equations $x = 2\cos\theta$, $y = 2\sin 2\theta$.
Obtain the cartesian equation of the curve in the form $y^2 = f(x)$.

Pattern parametrics

You can produce many interesting curves using parametric equations.
Most of those considered here go beyond what you will meet in the exam.

The one shown here uses only simple polynomials:

$x = t^4 - 4t^2$, $y = t^3 - 3t$ ($t = -3$ to 3)

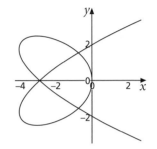

(It gets its swooping character because x in terms of t has three
stationary points within the stated limits of t, while y in terms of t
has two stationary points. Bear this in mind if you want to want
to produce something similar of your own.)

Other curves besides a circle or ellipse – such as this one – can be
produced using sine and cosine. Try to work out what equations
have been used, then check with a graph plotter.

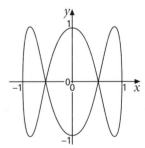

Plot some of these on a graph plotter, which must be set to degrees.
Conjecture what effect a particular modification (such as changing a constant)
will have on the curve, then test and see if you were right.

1 $x = \cos 5t$, $y = \sin 3t$ ($t = 0$ to 400)

2 $x = \sin t - \cos 2t$, $y = \cos t - \sin 2t$ ($t = 0$ to 360) (trefoil)

3 $x = \dfrac{t^3}{50}$, $y = \sin 100t$ ($t = -10$ to 10)

4 $x = \cos t + \cos^2 t$, $y = \sin t + \cos t \sin t$ ($t = 0$ to 360) (cardioid)

5 $x = 3(1 - 0.01t)\cos 99t$, $y = 3(1 - 0.01t)\sin 100t$ ($t = 0$ to 95)

6 $x = 3\sin t + \sin 12t$, $y = 3\cos t + \cos 12t$ ($t = 0$ to 360)

7 $x = at + \cos 100t$ (where $a = 0.2, 0.5, 1, 2$), $y = \sin 100t$ ($t = -10$ to 10)

8 $x = 3\sin t + \sin 10t$, $y = 3\cos t + \cos 13t$ ($t = 0$ to 360)

Key points

- Two functions that separately define the x- and y-coordinates of a graph in terms of a third variable are called parametric equations.

  The third variable is called the parameter. (p 18)

- For a graph defined parametrically, transformations are applied as follows:

  To apply a translation $\begin{bmatrix} a \\ b \end{bmatrix}$ add a to the x-function and b to the y-function.

  (This may be used to give a circle or ellipse a centre other than the origin.)

  To stretch in the x-direction, multiply the x-function by the required factor; correspondingly for a stretch in the y-direction.

  To reflect in the y-axis multiply the x-function by -1; to reflect in the x-axis multiply the y-function by -1. (p 19)

- To convert a pair of parametric equations to a single cartesian equation, eliminate the parameter using methods like those used for simultaneous equations. (pp 21–23)

- Where parametric equations involve trigonometric functions, the parameter can often be eliminated by substituting into an appropriate trigonometric identity. (pp 31–32)

- The curve $x = r\cos\theta$, $y = r\sin\theta$ is a circle with radius r, centre the origin. (p 28)

- The curve $x = a\cos\theta$, $y = b\sin\theta$ is an ellipse, centre the origin.
  Its width is $2a$ units and its height is $2b$ units.

  Its cartesian equation is $\dfrac{x^2}{a^2} + \dfrac{y^2}{b^2} = 1$. (pp 28–29)

Test yourself (answers p 160)

1 A curve is defined by $x = \dfrac{1}{t^2}$, $y = 3t$ $(t \neq 0)$.

Find the coordinates of the points on the curve where $t = -3, -2, -1, 1, 2$ and 3.

2 Find the coordinates of the point(s) where each of the following curves meets the x- and y-axes.

(a) $x = t - 3$, $y = 3t + 1$

(b) $x = t^2 - 2t$, $y = 5 - t$

(c) $x = 2t - 8$, $y = t^2 + t - 6$

(d) $x = \dfrac{t+1}{t-1}$, $y = t^3 + 27$ $(t \neq 1)$

3 Give the coordinates of the points where the curve defined by $x = t^2 - 5t$, $y = \dfrac{t+1}{t}$ meets the line $x = -6$.

4 Convert each of these pairs of parametric equations to an equation of the form $ax + by + c = 0$.

(a) $x = 4t + 5$, $y = 3t$

(b) $x = 5 - 2t$, $y = \frac{1}{3}t$

5 Obtain a single cartesian equation for each pair of parametric equations.

(a) $x = 3t, \; y = t^2$

(b) $x = t^3, \; y = \frac{1}{2}t$

(c) $x = \sqrt{t}, \; y = t^3 \; (t \geq 0)$

(d) $x = t - 1, \; y = 2 - t^2$

(e) $x = \dfrac{1}{t}, \; y = 3t \; (t \neq 0)$

(f) $x = t^2 - t, \; y = \frac{1}{4}t$

(g) $x = t + 2, \; y = t^3 + 2t$

(h) $x = \dfrac{1}{2t}, \; y = 3 - t \; (t \neq 0)$

(i) $x = 3t^3 + 2, \; y = 5 - t^3$

(j) $x = t(3 - t), \; y = \sqrt{t} \; (t \geq 0)$

(k) $x = t^2, \; y = \dfrac{1}{t - 2} \; (t \neq 2)$

(l) $x = \dfrac{t}{t - 1}, \; y = \dfrac{t}{t + 1} \; (t \neq -1, t \neq 1)$

(m) $x = 3\sqrt{t} - 2, \; y = 4 - \sqrt{t} \; (t \geq 0)$

(n) $x = \dfrac{t}{2t - 3}, \; y = \dfrac{t}{t + 1} \; (t \neq \frac{3}{2}, t \neq -1)$

6 A curve is defined parametrically by $x = t^2 + \dfrac{2}{t}, \; y = t^2 - \dfrac{2}{t} \; (t \neq 0)$.

By first expressing $x + y$ and $x - y$ in terms of t, find the cartesian equation.

7 Give a pair of parametric equations for each of these.

(a)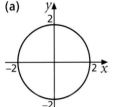

(b)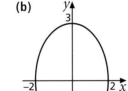

(c)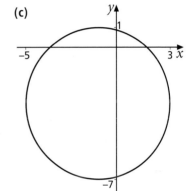

(d)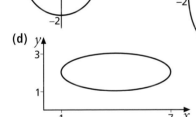

8 Find a pair of parametric equations for each of these ellipses.

(a) $\dfrac{x^2}{4} + \dfrac{y^2}{16} = 1$

(b) $25x^2 + 4y^2 = 100$

9 An ellipse is defined by the parametric equations $x = 6\cos\theta, \; y = 2\sin\theta$.
A chord of the ellipse has its ends at points on the ellipse where $\theta = \frac{1}{6}\pi$ and $\frac{5}{6}\pi$
respectively. Find the length of the chord, giving your answer as an exact value.

10 Obtain a single cartesian equation for each pair of parametric equations.

(a) $x = 3\cos\theta, \; y = 4\sin\theta$

(b) $x = 3\cos\theta + 5, \; y = 3\sin\theta - 2$

(c) $x = 2\cos\theta, \; y = \sec\theta$

(d) $x = \operatorname{cosec}\theta + 1, \; y = \sin^2\theta$

(e) $x = 3\tan\theta, \; y = 4 - 2\sec\theta$

(f) $x = 4\cos 2\theta - 1, \; y = \sin\theta + 2$

(g) $x = \sin\theta, \; y = \tan\theta$

(h) $x = \cos\theta + 3, \; y = \sin 2\theta \; \left(0 \leq \theta \leq \dfrac{\pi}{2}\right)$

3 The binomial theorem

In this chapter you will learn how to use the binomial theorem for negative and fractional indices

A Reviewing the binomial theorem for positive integers

In Core 2, you expanded expressions such as $(1 - 2x)^4$ using the binomial theorem.
Using the theorem to expand $(1 + ax)^n$ gives

$$(1 + ax)^n = 1 + nax + \frac{n(n-1)}{2!}(ax)^2 + \frac{n(n-1)(n-2)}{3!}(ax)^3 + \ldots + \frac{n(n-1)(n-2)\ldots 1}{n!}(ax)^n$$

So $(1 - 2x)^4 = (1 + (-2x))^4$

$$= 1 + 4 \times (-2x) + \frac{4 \times 3}{2!} \times (-2x)^2 + \frac{4 \times 3 \times 2}{3!} \times (-2x)^3 + \frac{4 \times 3 \times 2 \times 1}{4!} \times (-2x)^4$$

$$= 1 - 8x + 24x^2 - 32x^3 + 16x^4$$

The expansion for $(1 + ax)^n$ can be used to expand an expression such as $(2 + 6x)^5$ as follows.
$$(2 + 6x)^5 = (2(1 + 3x))^5 = 2^5(1 + 3x)^5$$

Now $(1 + 3x)^5 = 1 + 5 \times (3x) + \frac{5 \times 4}{2!} \times (3x)^2 + \frac{5 \times 4 \times 3}{3!} \times (3x)^3 +$

$$\frac{5 \times 4 \times 3 \times 2}{4!} \times (3x)^4 + \frac{5 \times 4 \times 3 \times 2 \times 1}{5!} \times (3x)^5$$

$$= 1 + 15x + 90x^2 + 270x^3 + 405x^4 + 243x^5$$

So $(2 + 6x)^5 = 2^5(1 + 3x)^5$

$$= 2^5(1 + 15x + 90x^2 + 270x^3 + 405x^4 + 243x^5)$$

$$= 32 + 480x + 2880x^2 + 8640x^3 + 12\,960x^4 + 7776x^5$$

Exercise A (answers p 161)

1 Use the binomial theorem to expand $(1 + 5x)^6$ completely.

2 Find the first five terms in the expansion of $(1 - 3x)^{12}$.

3 Find the first three terms in the expansion of $(4 + 8x)^7$.

4 (a) Find the first four terms in the expansion of $\left(1 + \frac{1}{2}x\right)^8$.
    (b) Hence find the first four terms in the expansion of $(2 + x)^8$.

5 (a) Expand $(3 + x)^6$ fully.
    (b) Use the expansion to find the value of 3.01^6 correct to two decimal places.

B Extending the binomial theorem (answers p 161)

New ideas and techniques often arise when mathematicians consider extending the scope of a definition or formula.

We know that the binomial theorem helps us expand expressions of the form $(1 + ax)^n$ where n is a positive integer.

Can we use the binomial theorem when n is a negative integer?

For example, consider the expression $(1 + x)^{-2}$.

Using the formula for the binomial theorem gives

$$(1 + x)^{-2} = 1 + (-2)x + \frac{(-2)(-3)}{2!}x^2 + \frac{(-2)(-3)(-4)}{3!}x^3 + \frac{(-2)(-3)(-4)(-5)}{4!}x^4 + \ldots$$

D

B1 (a) Show that the expansion above simplifies to
$$(1 + x)^{-2} = 1 - 2x + 3x^2 - 4x^3 + 5x^4 - \ldots$$

(b) Find and simplify the next three terms in this expansion.

(c) Why will the formula let us produce as many terms as we like when we apply it to $(1 + x)^{-2}$ but only produces a finite number of terms for, say, $(1 + x)^2$?

(d) What will be the 15th term in the expansion for $(1 + x)^{-2}$?

How is the expansion $1 - 2x + 3x^2 - 4x^3 + 5x^4 - \ldots$ related to $(1 + x)^{-2}$?
We can try to answer this question by comparing $1 - 2x + 3x^2 - 4x^3 + 5x^4 - \ldots$
with $(1 + x)^{-2}$ for different values of x.

B2 (a) (i) Use your calculator to find the value of $(1 + x)^{-2}$ when $x = 0.2$, correct to 6 d.p.

(ii) Copy and complete the following table for the expansion
$1 - 2x + 3x^2 - 4x^3 + 5x^4 - \ldots$ with $x = 0.2$.

Number of terms	Expansion	Value when $x = 0.2$
1	1	1
2	$1 - 2x$	0.6
3	$1 - 2x + 3x^2$	
4	$1 - 2x + 3x^2 - 4x^3$	
5		
6		
7		
8		

(iii) What appears to be happening?

(b) (i) Work out the value of $(1 + x)^{-2}$ when $x = -2$.

(ii) In your table, complete a column for $x = -2$ and comment on your results.

(c) (i) Use your calculator to find the value of $(1 + x)^{-2}$ when $x = 0.9$, correct to 6 d.p.

(ii) In your table, complete a column for $x = 0.9$ and comment on your results.

With an infinite series such as $1 - 2x + 3x^2 - 4x^3 + 5x^4 - \ldots$ you may need to evaluate the sum of many terms before you can see whether or not the series is converging to a limit.

A spreadsheet is useful for this.

B3 The spreadsheet below is set up to investigate further the value of the expansion $1 - 2x + 3x^2 - 4x^3 + 5x^4 - \ldots$ when $x = 0.9$.

The formula used here is '=(−1)^(A2+1)*A2' and it is filled down.

This is the value of x.

'1' is entered in cell C2.

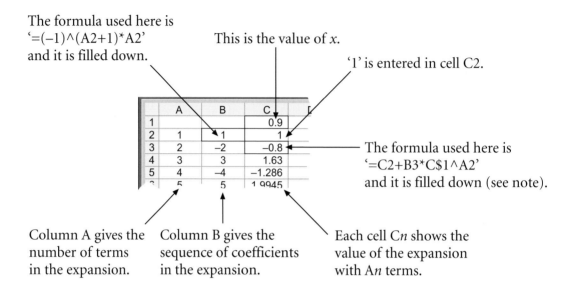

	A	B	C	
1			0.9	
2	1	1	1	
3	2	−2	−0.8	
4	3	3	1.63	
5	4	−4	−1.286	
	5	5	1.9945	

The formula used here is '=C2+B3*C$1^A2' and it is filled down (see note).

Column A gives the number of terms in the expansion.

Column B gives the sequence of coefficients in the expansion.

Each cell Cn shows the value of the expansion with An terms.

Note: 'C$1' ensures that cell C1 is always referred to (not 'the cell two rows above') when the formula is filled down.

(a) Set up a spreadsheet as above.
    Do you think the expansion converges when $x = 0.9$?

(b) Try some different values of x, positive and negative, in cell C1.
    Can you find a rule that tells you when $1 - 2x + 3x^2 - 4x^3 + 5x^4 - \ldots$ will converge?

B4 Define functions f, g and h as
$$f(x) = (1 + x)^{-2}$$
$$g(x) = 1 - 2x + 3x^2 - 4x^3 + 5x^4 - 6x^5$$
$$h(x) = 1 - 2x + 3x^2 - 4x^3 + 5x^4 - 6x^5 + 7x^6 - 8x^7 + 9x^8 - 10x^9 + 11x^{10} - 12x^{11}$$

Use a graphic calculator or graph plotter to draw the graphs of $y = f(x)$, $y = g(x)$ and $y = h(x)$ for $-2 \leq x \leq 2$.

What do you notice about your graphs?

It can be proved (using methods outside the scope of this book) that the series expansion $1 - 2x + 3x^2 - 4x^3 + 5x^4 - \dots$ converges to $(1 + x)^{-2}$ for values of x such that $-1 < x < 1$. This can be written as $|x| < 1$.

We say that the expansion $(1 + x)^{-2} = 1 - 2x + 3x^2 - 4x^3 + 5x^4 - \dots$ is **valid** for $|x| < 1$.

B5 (a) Show that if the binomial theorem is used to find an expansion for $(1 - 2x)^{-1}$ then the first six terms are $1 + 2x + 4x^2 + 8x^3 + 16x^4 + 32x^5$.

(b) What will be the 12th term in this expansion for $(1 - 2x)^{-1}$?

(c) (i) Find the value of $(1 - 2x)^{-1}$ when $x = 0.7$.

(ii) Edit the spreadsheet used in B3 to investigate the value of the expansion $1 + 2x + 4x^2 + 8x^3 + 16x^4 + 32x^5 + \dots$ when $x = 0.7$.

(d) Try some different values of x, positive and negative. Can you find a rule that tells you when $1 + 2x + 4x^2 + 8x^3 + 16x^4 + 32x^5 + \dots$ will converge?

K The binomial theorem can produce an expansion for $(1 + ax)^n$ when n is negative but it is only valid for certain values for x, namely the set of values for which $-1 < ax < 1$.

Will there be a similar result if n is a fraction?

For example, applying the formula to $(1 + x)^{\frac{1}{2}}$ gives

$$(1 + x)^{\frac{1}{2}} = 1 + \left(\tfrac{1}{2}\right)x + \frac{\left(\frac{1}{2}\right)\left(-\frac{1}{2}\right)}{2!}x^2 + \frac{\left(\frac{1}{2}\right)\left(-\frac{1}{2}\right)\left(-\frac{3}{2}\right)}{3!}x^3 + \dots$$

B6 (a) Show that the expansion above simplifies to $(1 + x)^{\frac{1}{2}} = 1 + \tfrac{1}{2}x - \tfrac{1}{8}x^2 + \tfrac{1}{16}x^3 + \dots$

(b) Find and simplify the next two terms in the expansion.

(c) When $x = 0.8$, find the value of

(i) $(1 + x)^{\frac{1}{2}}$ **(ii)** $1 + \tfrac{1}{2}x - \tfrac{1}{8}x^2 + \tfrac{1}{16}x^3$

(d) Use a graphic calculator or graph plotter to show that the graph of $y = 1 + \tfrac{1}{2}x - \tfrac{1}{8}x^2 + \tfrac{1}{16}x^3$ is very close to the graph of $y = (1 + x)^{\frac{1}{2}}$ when $|x| < 1$.

It was Isaac Newton who discovered the following.

K The **binomial expansion**

$$(1 + ax)^n = 1 + nax + \frac{n(n-1)}{2!}(ax)^2 + \frac{n(n-1)(n-2)}{3!}(ax)^3 + \dots$$

holds for **all** negative and fractional values of n provided that $-1 < ax < 1$, that is $|ax| < 1$.

Newton's work on infinite series led directly to the development of calculus. The binomial expansion is still an important tool today in many areas of advanced mathematics.

To work out when a binomial expansion is valid you need to be able to solve inequalities that involve the modulus function.

B7 Show that

(a) $|2x| < 1$ is equivalent to $|x| < \tfrac{1}{2}$ **(b)** $|-4x| < 1$ is equivalent to $|x| < \tfrac{1}{4}$

(c) $|\tfrac{1}{3}x| < 1$ is equivalent to $|x| < 3$ **(d)** $|\tfrac{3}{4}x| < 1$ is equivalent to $|x| < \tfrac{4}{3}$

Example 1

Find the binomial expansion of $\dfrac{1}{(1+3x)^2}$ in ascending powers of x as far as the term in x^3 and state the set of values of x for which the expansion is valid.

Solution

$$\frac{1}{(1+3x)^2} = (1+3x)^{-2} = 1 + (-2)(3x) + \frac{(-2)(-3)}{2!}(3x)^2 + \frac{(-2)(-3)(-4)}{3!}(3x)^3 + \dots$$

$$\approx 1 - 6x + 27x^2 - 108x^3$$

The expansion is valid when $-1 < 3x < 1$.

This simplifies to $-\frac{1}{3} < x < \frac{1}{3}$, or $|x| < \frac{1}{3}$.

Example 2

Expand $(1-2x)^{\frac{1}{2}}$ in ascending powers of x as far as the term in x^3.
Show that you can use the expansion to find an approximate value for $\sqrt{0.94}$.
Hence find $\sqrt{0.94}$ correct to two decimal places.

Solution

$$(1-2x)^{\frac{1}{2}} = 1 + \left(\tfrac{1}{2}\right)(-2x) + \frac{\left(\tfrac{1}{2}\right)\left(-\tfrac{1}{2}\right)}{2!}(-2x)^2 + \frac{\left(\tfrac{1}{2}\right)\left(-\tfrac{1}{2}\right)\left(-\tfrac{3}{2}\right)}{3!}(-2x)^3 + \dots$$

$$\approx 1 - x - \tfrac{1}{2}x^2 - \tfrac{1}{2}x^3$$

The expansion is valid when $|-2x| < 1$, which gives $|x| < \frac{1}{2}$.

$\sqrt{0.94} = 0.94^{\frac{1}{2}} = (1 - 2 \times 0.03)^{\frac{1}{2}}$, which can be obtained by substituting $x = 0.03$ into $(1-2x)^{\frac{1}{2}}$.

As $|0.03| < \frac{1}{2}$, the expansion is valid for this value.

So $\sqrt{0.94} = (1 - 2 \times 0.03)^{\frac{1}{2}} \approx 1 - (0.03) - \tfrac{1}{2}(0.03)^2 - \tfrac{1}{2}(0.03)^3 = 1 - 0.03 - 0.00045 - 0.0000135$

$$= 0.9695365$$

Further terms in the expansion will not change the first three decimal places,
so we have $\sqrt{0.94} = 0.97$ correct to two decimal places.

Example 3

Find the series expansion of $(3+x)^{-1}$ as far as the term in x^3 and state the set of values of x for which the expansion is valid.

Solution

$$(3+x)^{-1} = \left(3\left(1 + \tfrac{1}{3}x\right)\right)^{-1} = 3^{-1}\left(1 + \tfrac{1}{3}x\right)^{-1} = \tfrac{1}{3}\left(1 + \tfrac{1}{3}x\right)^{-1}$$

$$= \tfrac{1}{3}\left(1 + (-1)\left(\tfrac{1}{3}x\right) + \frac{(-1)(-2)}{2!}\left(\tfrac{1}{3}x\right)^2 + \frac{(-1)(-2)(-3)}{3!}\left(\tfrac{1}{3}x\right)^3 + \dots\right)$$

$$\approx \tfrac{1}{3} - \tfrac{1}{9}x + \tfrac{1}{27}x^2 - \tfrac{1}{81}x^3$$

The expansion is valid when $\left|\tfrac{1}{3}x\right| < 1$, which gives $|x| < 3$.

Example 4

Expand $(1 + x)^{\frac{2}{3}}$ in ascending powers of x as far as the term in x^2.

Use the expansion to find an approximate value for $1.6^{\frac{2}{3}}$ and comment on its accuracy.

Solution

$(1 + x)^{\frac{2}{3}} = 1 + \left(\frac{2}{3}\right)x + \dfrac{\left(\frac{2}{3}\right)\left(-\frac{1}{3}\right)}{2!}x^2 + \ldots \approx 1 + \frac{2}{3}x - \frac{1}{9}x^2$

So $1.6^{\frac{2}{3}} = (1 + 0.6)^{\frac{2}{3}} \approx 1 + \frac{2}{3}\times0.6 - \frac{1}{9}\times(0.6)^2$ $|0.6| < 1$ *so we can use the expansion.*

$\qquad\qquad = 1 + 0.4 - 0.04$

$\qquad\qquad = 1.36$

Using a calculator gives $1.6^{\frac{2}{3}} = 1.367\,980\,757\ldots$ which agrees with the approximate value when both values are rounded to two significant figures (to give 1.4). So the approximate value is accurate to two significant figures.

Exercise B (answers p 162)

1 Expand each of these as a series of ascending powers of x as far as the term in x^3.
For each expansion, give the range of values, in the form $|x| < k$, for which it is valid.

   (a) $(1 + x)^{-1}$ **(b)** $(1 - x)^{-2}$ **(c)** $(1 + 2x)^{-4}$

   (d) $\left(1 + \frac{1}{2}x\right)^{-2}$ **(e)** $\dfrac{1}{1 + 3x}$ **(f)** $\dfrac{1}{\left(1 - \frac{1}{3}x\right)^2}$

   (g) $(1 - x)^{\frac{1}{2}}$ **(h)** $\sqrt{1 + \frac{1}{4}x}$ **(i)** $\dfrac{1}{\sqrt{1 + 2x}}$

2 (a) Show that the first four terms of the binomial expansion for $(1 + 3x)^{\frac{1}{3}}$ are
$$1 + x - x^2 + \tfrac{5}{3}x^3$$
   (b) State the range of values for which the expansion is valid.

3 (a) Obtain the binomial expansion of $(1 + 4x)^{\frac{1}{2}}$ as far as the term in x^6.

   (b) Show that this expansion can be used to find an approximation for $\sqrt{1.4}$.
Hence find the value of $\sqrt{1.4}$ correct to two decimal places.

4 (a) Obtain the binomial expansion of $(1 + 2x)^{\frac{3}{2}}$ as far as the term in x^2.

   (b) Use this expansion to find an approximate value for $1.8^{\frac{3}{2}}$.
Comment on its accuracy.

5 (a) Obtain the binomial expansion of $\left(1 + \frac{1}{2}x\right)^{\frac{1}{2}}$ as far as the term in x^2.

   (b) (i) Hence find the series expansion of $(16 + 8x)^{\frac{1}{2}}$ as far as the term in x^2.

        (ii) Find the range of values of x for which this expansion is valid.

6 Find the coefficient of x^2 in the series expansion of $\dfrac{1}{\sqrt{9 - x}}$.

7 Show that the coefficient of x^3 in the series expansion of $(8 - 3x)^{-\frac{1}{3}}$ is $\frac{7}{1536}$.

8 Expand each of these as a series of ascending powers of x as far as the term in x^2.
In each case give the range of values for which the expansion is valid.

(a) $(2 + x)^{-1}$

(b) $\dfrac{1}{(4 - x)^3}$

(c) $(6 + 3x)^{-2}$

(d) $(4 + x)^{\frac{1}{2}}$

(e) $(3 + 2x)^{-1}$

(f) $\sqrt[5]{32 - 64x}$

C Multiplying to obtain expansions

Sometimes a series expansion can be obtained for a more complicated rational expression by finding expansions for simpler rational expressions and then multiplying.

Example 5

Expand $\dfrac{1 + 3x}{(1 + x)^3}$, $|x| < 1$, in ascending powers of x as far as the term in x^3.

Solution

$$\frac{1 + 3x}{(1 + x)^3} = (1 + 3x)(1 + x)^{-3}$$

$$= (1 + 3x)\left(1 + (-3)x + \frac{(-3)(-4)}{2!}x^2 + \frac{(-3)(-4)(-5)}{3!}x^3 + \cdots\right) \quad \textit{You only need to go as far as the term in } x^3.$$

$$= (1 + 3x)(1 - 3x + 6x^2 - 10x^3 + \cdots)$$

$$= 1 - 3x + 6x^2 - 10x^3 + \cdots + 3x - 9x^2 + 18x^3 - \cdots \quad \textit{Multiply as far as terms in } x^3.$$

$$\approx 1 - 3x^2 + 8x^3$$

Exercise C (answers p 163)

1 For each of these, find an expansion in ascending powers of x as far as the term in x^3.

(a) $\dfrac{5}{(1 - x)^2}$, $|x| < 1$

(b) $\dfrac{x}{1 + 3x}$, $|x| < \frac{1}{3}$

(c) $\dfrac{x + 1}{\sqrt{1 - 4x}}$, $|x| < \frac{1}{4}$

(d) $\dfrac{1 - 2x}{(1 - 3x)^3}$, $|x| < \frac{1}{3}$

2 (a) (i) Show that the first three terms of the expansion for $\dfrac{1}{1 - 3x}$ are $1 + 3x + 9x^2$.

(ii) State the range of values for which the expansion is valid.

(b) (i) Show that the first three terms of the expansion for $\dfrac{1}{(1 + x)^4}$ are $1 - 4x + 10x^2$.

(ii) State the range of values for which the expansion is valid.

(c) (i) By considering the product $(1 + 3x + 9x^2 + \cdots)(1 - 4x + 10x^2 - \cdots)$, find an expansion for $\dfrac{1}{(1 - 3x)(1 + x)^4}$ in ascending powers of x as far as the term in x^2.

(ii) Explain why this expansion is only valid for $|x| < \frac{1}{3}$.

D Adding (using partial fractions) to obtain expansions

An expansion for an expression such as $\dfrac{10-x}{(3-x)(1+2x)}$ can be obtained by finding expansions for simpler rational expressions and then multiplying, but the product is quite complicated. An alternative method is to use partial fractions and then **add** the resulting expansions.

Example 6

The function f is given by $f(x) = \dfrac{10-x}{(3-x)(1+2x)}$.

Express $f(x)$ in partial fractions and hence obtain the expansion of $f(x)$ as far as the term in x^2. State the range of values for which the expansion is valid.

Solution

Let
$$\frac{10-x}{(3-x)(1+2x)} \equiv \frac{A}{3-x} + \frac{B}{1+2x}$$

Then
$$\frac{10-x}{(3-x)(1+2x)} \equiv \frac{A(1+2x)+B(3-x)}{(3-x)(1+2x)}$$

giving
$$A(1+2x) + B(3-x) \equiv 10-x$$

Substituting $x = 3$ in this identity gives $\quad 7A = 7$, which gives $A = 1$.

Substituting $x = -\frac{1}{2}$ in this identity gives $\quad \frac{7}{2}B = \frac{21}{2}$, which gives $B = 3$.

Hence
$$f(x) \equiv \frac{1}{3-x} + \frac{3}{1+2x}$$

Now, $\dfrac{1}{3-x} = (3-x)^{-1} = 3^{-1}\left(1 - \frac{1}{3}x\right)^{-1} = \frac{1}{3}\left(1 - \frac{1}{3}x\right)^{-1}$

$$= \frac{1}{3}\left(1 + (-1)\left(-\frac{1}{3}x\right) + \frac{(-1)(-2)}{2!}\left(-\frac{1}{3}x\right)^2 + \dots\right)$$

$$= \frac{1}{3}\left(1 + \frac{1}{3}x + \frac{1}{9}x^2 + \dots\right)$$

$$= \frac{1}{3} + \frac{1}{9}x + \frac{1}{27}x^2 + \dots$$

Also, $\dfrac{3}{1+2x} = 3(1+2x)^{-1} = 3\left(1 + (-1)(2x) + \frac{(-1)(-2)}{2!}(2x)^2 + \dots\right)$

$$= 3(1 - 2x + 4x^2 - \dots)$$

$$= 3 - 6x + 12x^2 - \dots$$

So $f(x) = \frac{1}{3} + \frac{1}{9}x + \frac{1}{27}x^2 + \dots + 3 - 6x + 12x^2 - \dots$

$$\approx \frac{10}{3} - \frac{53}{9}x + \frac{325}{27}x^2$$

The expansion for $\left(1 - \frac{1}{3}x\right)^{-1}$ is valid for $\left|-\frac{1}{3}x\right| < 1$, which gives $|x| < 3$.

The expansion for $(1 + 2x)^{-1}$ is valid for $|2x| < 1$, which gives $|x| < \frac{1}{2}$.

The expansion for $f(x)$ is valid for the values that satisfy both inequalities, that is for $|x| < \frac{1}{2}$.

Exercise D (answers p 163)

1 The function f is given by $f(x) = \dfrac{4 + 5x}{(1 - x)(1 + 2x)}$.

    (a) Express $f(x)$ as partial fractions.

    (b) Hence obtain the expansion of $f(x)$ in ascending powers of x as far as the term in x^2.

    (c) Show that the expansion is valid when $|x| < \frac{1}{2}$.

2 The function g is given by $g(x) = \dfrac{8 - x}{(2 + x)(1 - 2x)}$.

    (a) Express $g(x)$ as partial fractions.

    (b) (i) Show that the first three terms in the expansion of $\dfrac{1}{2 + x}$ in ascending powers of x are $\frac{1}{2} - \frac{1}{4}x + \frac{1}{8}x^2$.

        (ii) Obtain a similar expansion for $\dfrac{1}{1 - 2x}$.

    (c) Hence obtain the first three terms in the expansion of $g(x)$ in ascending powers of x.

    (d) Find the range of values of x for which this expansion of $g(x)$ is valid.

3 Find the first four terms of the series expansion for each of these rational expressions. Find the range of values for which each expansion is valid.

    (a) $\dfrac{2 + x}{(1 + 4x)(1 - 3x)}$ **(b)** $\dfrac{7 + x}{(3 - x)(1 + 3x)}$ **(c)** $\dfrac{9 - x}{(1 + x)(3 - 2x)}$

4 Show that the first three terms in the expansion of $\dfrac{11x - 3}{(4 - 3x)(1 + x)}$ are $-\frac{3}{4} + \frac{47}{16}x - \frac{83}{64}x^2$.

Key points

- The binomial expansion
$$(1 + ax)^n = 1 + nax + \frac{n(n-1)}{2!}(ax)^2 + \frac{n(n-1)(n-2)}{3!}(ax)^3 + \ldots$$
is valid for negative and fractional values of n for $|ax| < 1$.   (p 39)

Mixed questions (answers p 164)

1 (a) Expand $\dfrac{4}{(1 + 2x)^3}$ in ascending powers of x as far as the term in x^2.

    (b) For what values of x is this expansion valid?

2 (a) Expand $(100 - 400x)^{\frac{1}{2}}$ in ascending powers of x up to and including the term in x^3.

    (b) (i) What value of x gives an approximate value for $\sqrt{60}$ when used in this expansion?

        (ii) Use the series expansion in part (a) to find an approximate value for $\sqrt{60}$.

        (iii) Comment on the accuracy of this approximate value.

3 In the expansion of $(1 + ax)^n$ the coefficients of x and x^2 are -10 and 75 respectively. Find the value of a and the value of n.

4 Show that, in the expansion of $(16 + 192x)^{\frac{3}{4}}$, the coefficient of x^3 is 540.

5 **(a)** Expand $(1 + x)^{-1}$ in ascending powers of x as far as the term in x^3.

(b) By integrating $(1 + x)^{-1}$ and its expansion, find a series expansion for $\ln(1 + x)$ as far as the term in x^4.

(c) State, with a reason, whether your series expansion in part (b) could be used to find an approximation for $\ln 3$.

6 **(a)** Obtain the first four terms in the expansion of $(1 - 2x)^{-1}$ in ascending powers of x.

(b) Hence show that, for small x, $\dfrac{1-x}{1-2x} \approx 1 + x + 2x^2 + 4x^3$.

(c) Taking a suitable value for x, which should be stated, use the series expansion in part (b) to find an approximate value for $\frac{99}{98}$, giving your answer correct to 5 d.p.

(d) Hence write down an approximate value for $\frac{1}{98}$.

Test yourself (answers p 165)

1 The binomial expansion of $(1 - 2x)^{-3}$ in ascending powers of x up to and including the term in x^3 is $1 + 6x + px^2 + qx^3$.

(a) Find the value of p and the value of q.

(b) **(i)** Hence obtain the first three terms of the expansion for $\dfrac{1-3x}{(1-2x)^3}$.

(ii) State the range of values for which this expansion is valid.

2 **(a)** Given that $\dfrac{2(15-13x)}{(1-x)(3+x)} \equiv \dfrac{A}{1-x} + \dfrac{B}{3+x}$, find the values of the constants A and B.

(b) Hence, or otherwise, find the series expansion in ascending powers of x, up to and including the term in x^3, of $\dfrac{2(15-13x)}{(1-x)(3+x)}$, for $|x| < 1$.

3 **(a)** Prove that, when $x = \frac{1}{15}$, the value of $(1 + 5x)^{-\frac{1}{2}}$ is exactly equal to $\sin 60°$.

(b) Expand $(1 + 5x)^{-\frac{1}{2}}$, $|x| < 0.2$ in ascending powers of x up to and including the term in x^3, simplifying each term.

(c) Use your answer to part (b) to find an approximation for $\sin 60°$.

(d) Find the difference between the exact value of $\sin 60°$ and the approximation in part (c).

Edexcel

4 **(a)** Expand $(1 - 10x)^{\frac{1}{5}}$, $|x| < \frac{1}{10}$, in ascending powers of x as far as the term in x^3.

(b) By substituting $x = 0.001$ in your expansion, find $\sqrt[5]{99000}$ correct to 5 s.f.

5 Find the coefficient of x^2 in the series expansion of $\dfrac{x+8}{\sqrt{4-x}}$.

4 Differentiation

In this chapter you will learn how to
- differentiate a function defined parametrically
- differentiate a function defined implicitly
- find the equation of the tangent and the normal to a curve defined parametrically or implicitly

---

Key points from previous books

- The line through the point (x_1, y_1) with gradient m has the equation $y - y_1 = m(x - x_1)$.

- A line perpendicular to a line with gradient m has gradient $-\dfrac{1}{m}$.

- The derivative of e^x is e^x.

- The derivative of $\ln x$ is $\dfrac{1}{x}$.

- $\dfrac{dy}{dx} = \dfrac{1}{\frac{dx}{dy}}$

- If $y = uv$, then $\dfrac{dy}{dx} = u\dfrac{dv}{dx} + v\dfrac{du}{dx}$ (product rule)

- If $y = \dfrac{u}{v}$, then $\dfrac{dy}{dx} = \dfrac{v\dfrac{du}{dx} - u\dfrac{dv}{dx}}{v^2}$ (quotient rule)

- The derivative of $\sin x$ is $\cos x$.
  The derivative of $\cos x$ is $-\sin x$.
  The derivative of $\tan x$ is $\sec^2 x$.

- $\dfrac{dy}{dx} = \dfrac{dy}{du} \times \dfrac{du}{dx}$ (chain rule)

- The derivative of $f(ax)$ is $af'(ax)$.
  The derivative of $f(ax + b)$ is $af'(ax + b)$.

A Functions defined parametrically (answers p 165)

In each of the equations below, y is defined **explicitly** in terms of x.

$$y = 3x^2 - 1 \qquad y = \sqrt{x+2} \ \ (x \geq -2) \qquad y = \ln x \ \ (x > 0) \qquad y = x^2 e^x$$

If x and y are each defined in terms of a third variable t, the third variable is called a **parameter**. The function that links y to x is said to be defined **parametrically**.

For example, the equations $x = 3t$ and $y = t^2 + 1$ together define a function connecting x and y, as the table of values shows.

t	-3	-2	-1	0	1	2	3
$x = 3t$	-9	-6	-3	0	3	6	9
$y = t^2 + 1$	10	5	2	1	2	5	10

The graph of y against x is shown here.
Points on the graph are labelled by the value of t.

The gradient at any point on this graph is the value
of $\dfrac{dy}{dx}$ at the point.

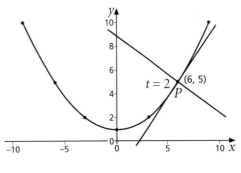

We can find an expression for $\dfrac{dy}{dx}$ **in terms of t** from
the two parametric equations.

The chain rule says that $\dfrac{dy}{dx} = \dfrac{dy}{dt} \times \dfrac{dt}{dx}$. It is also

true that $\dfrac{dt}{dx} = \dfrac{1}{\dfrac{dx}{dt}}$.

It follows that

$$\frac{dy}{dx} = \frac{\dfrac{dy}{dt}}{\dfrac{dx}{dt}}$$

For the example above, $\dfrac{dy}{dt} = 2t$ and $\dfrac{dx}{dt} = 3$, so $\dfrac{dy}{dx} = \dfrac{2t}{3}$.

A1 Given that $x = t^2$ and $y = t^3$, find $\dfrac{dy}{dx}$ in terms of t.

A2 Find $\dfrac{dy}{dx}$ in terms of t given that

   (a) $x = 2t$, $y = t^2 + t$ (b) $x = 3t$, $y = \dfrac{1}{t}$ $(t \neq 0)$ (c) $x = \cos t$, $y = \sin t$

The diagram shows again the graph whose parametric
equations are

   $x = 3t$, $y = t^2 + 1$

P is the point on this graph where $t = 2$.
The coordinates of P are $(6, 5)$.

The gradient of the tangent, that is $\dfrac{dy}{dx}$, is $\dfrac{2t}{3}$ (as shown

above). So at P the gradient of the tangent is $\dfrac{2 \times 2}{3} = \dfrac{4}{3}$.

To find the equation of the tangent at P, we use the fact that the equation of
the line through (x_1, y_1) with gradient m is $y - y_1 = m(x - x_1)$.

A3 Show that the equation of the tangent at P is
   $y = \frac{4}{3}x - 3$

The normal at P is perpendicular to the tangent, so its gradient is $-\dfrac{1}{\frac{4}{3}} = -\dfrac{3}{4}$.

A4 Show that the equation of the normal is
   $y = -\frac{3}{4}x + \frac{19}{2}$

A5 A curve is given by the parametric equations $x = 2t$, $y = \dfrac{1}{t}$ $(t \neq 0)$.

(a) Find $\dfrac{dy}{dx}$ in terms of t.

(b) What are the coordinates of the point on the curve where $t = 4$?

(c) Show that the equation of the tangent at the point where $t = 4$ is
$$y = -\tfrac{1}{32}x + \tfrac{1}{2}$$

(d) Rewrite this equation in the form $ax + by = c$, where a, b and c are integers.

A6 A curve is given by the parametric equations $x = \sin t$, $y = 1 - \cos t$.

(a) Show that $\dfrac{dy}{dx} = \tan t$.

(b) Find the equation of the normal to the curve at the point where $t = \dfrac{\pi}{4}$.

Example 1

A curve is given by the parametric equations $x = t + \dfrac{1}{t}$, $y = t - \dfrac{1}{t}$ $(t > 0)$.

(a) Find the equation of the tangent to the curve at the point where $t = 2$, in the form $ax + by = c$, where a, b and c are integers.

(b) Find the equation of the normal to the curve at the point where $t = 2$, in the form $ax + by = c$, where a, b and c are integers.

Solution

$$x = t + \frac{1}{t} = t + t^{-1} \qquad y = t - \frac{1}{t} = t - t^{-1}$$

$$\frac{dx}{dt} = 1 - \frac{1}{t^2} \qquad \frac{dy}{dt} = 1 + \frac{1}{t^2} \qquad \frac{dy}{dx} = \frac{\dfrac{dy}{dt}}{\dfrac{dx}{dt}} = \frac{1 + \dfrac{1}{t^2}}{1 - \dfrac{1}{t^2}} \qquad \text{When } t = 2, \ \frac{dy}{dx} = \frac{1 + \frac{1}{4}}{1 - \frac{1}{4}} = \frac{\frac{5}{4}}{\frac{3}{4}} = \frac{5}{3}$$

The coordinates of the point where $t = 2$ are $\left(2 + \frac{1}{2}, 2 - \frac{1}{2}\right)$, or $\left(\frac{5}{2}, \frac{3}{2}\right)$.

(a) The gradient of the tangent is $\frac{5}{3}$, so the equation is $\qquad y - \frac{3}{2} = \frac{5}{3}\left(x - \frac{5}{2}\right)$

Expand the brackets. $\qquad\qquad\qquad\qquad\qquad\qquad\qquad\qquad y - \frac{3}{2} = \frac{5}{3}x - \frac{25}{6}$

Multiply through by 6. $\qquad\qquad\qquad\qquad\qquad\qquad\qquad\quad 6y - 9 = 10x - 25$

Rearrange. $\qquad\qquad\qquad\qquad\qquad\qquad\qquad\qquad\qquad\quad 10x - 6y = 16$

$\qquad\qquad\qquad\qquad\qquad\qquad\qquad\qquad\qquad\qquad\qquad \Rightarrow \ 5x - 3y = 8$

(b) The gradient of the normal is $-\frac{3}{5}$, so the equation is $\qquad y - \frac{3}{2} = -\frac{3}{5}\left(x - \frac{5}{2}\right)$

Expand the brackets. $\qquad\qquad\qquad\qquad\qquad\qquad\qquad\qquad y - \frac{3}{2} = -\frac{3}{5}x + \frac{3}{2}$

Multiply through by 10. $\qquad\qquad\qquad\qquad\qquad\qquad\qquad 10y - 15 = -6x + 15$

Rearrange. $\qquad\qquad\qquad\qquad\qquad\qquad\qquad\qquad\qquad\quad 6x + 10y = 30$

$\qquad\qquad\qquad\qquad\qquad\qquad\qquad\qquad\qquad\qquad\qquad \Rightarrow \ 3x + 5y = 15$

Exercise A (answers p 165)

1 A curve is defined by the parametric equations $x = t^2$, $y = t^3$.

   (a) Show that $\dfrac{dy}{dx} = \frac{3}{2}t$.

   (b) A is the point on the curve where $t = 4$. Find

      (i) the coordinates of A

      (ii) the equation of the tangent at A

      (iii) the equation of the normal at A

2 A curve is defined by the parametric equations $x = \dfrac{2}{t} + 1$, $y = \sqrt{t} + 1$ $(t > 0)$.

   (a) Find $\dfrac{dy}{dx}$ in terms of t.

   (b) Show that the equation of the tangent at the point where $t = 1$ is $y = -\frac{1}{4}x + \frac{11}{4}$.

   (c) Find the equation of the normal at the point where $t = 1$, in the form
$y = ax + b$.

3 A curve is defined by the parametric equations $x = \cos\theta$, $y = 2\sin\theta$ $(0 \leq \theta \leq 2\pi)$.

   (a) Show that $\dfrac{dy}{dx} = -\dfrac{2}{\tan\theta}$.

   (b) A is the point on the curve for which $\theta = \dfrac{\pi}{4}$. State the exact values of the
coordinates of A.

   (c) Show that the equation of the tangent at A is $2x + y = 2\sqrt{2}$.

4 A curve is given parametrically by the equations $x = 3\cos t$, $y = 4\sin t - 1$ $(0 \leq t < 2\pi)$.
P is the point on the curve where $t = \frac{1}{3}\pi$.
Show that the normal to the curve at P has the equation $8y = 6\sqrt{3}x + 7\sqrt{3} - 8$.

5 A curve is defined by the parametric equations $x = t - \dfrac{1}{t}$, $y = t^2$ $(t > 0)$.

   (a) Find $\dfrac{dy}{dx}$ in terms of t.

   (b) A is the point on the curve for which $t = \frac{1}{2}$.
Find, in the form $ax + by + c = 0$, where a, b and c are integers, the equation of

      (i) the tangent to the curve at A

      (ii) the normal to the curve at A

6 A curve is given by the parametric equations $x = 2\cos^2\theta$, $y = \sin 2\theta$ $(0 \leq \theta < \pi)$.

   (a) Show that $\dfrac{dy}{dx} = -\cot 2\theta$.

   (b) Show that the equation of the tangent to the curve at the point
where $\theta = \dfrac{\pi}{6}$ is $\sqrt{3}y + x = 3$.

7 A curve is given parametrically by the equation $x = \cos^2 t$, $y = 3\sin t$ $(0 \leq t < 2\pi)$.
Find the equation of the tangent to the curve at the point where $t = \pi$.

B Functions defined implicitly (answers p 166)

The equation $xy + y = 9$ is not written in the explicit form $y = $ a function of x.

However, for a given value of x it is possible to find a corresponding value of y. For example, if $x = 2$, the equation becomes $2y + y = 9$, from which $3y = 9$ and so $y = 3$.

We say that in the equation $xy + y = 9$, y is defined **implicitly** as a function of x.

By finding y for different values of x we can plot the graph of the equation $xy + y = 9$.

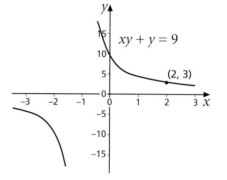

Other examples of implicit equations are

$$xy + x + y = 20$$
$$x + y + y^2 = 20$$
$$x^2 - y^2 = 1$$
$$x^2y + xy^2 = 16$$

In order to find the gradient function for a graph defined like this we need to be able to differentiate, with respect to x, expressions containing both x and y, such as $x + y + y^2$. Before we do this, some notation is needed.

In the symbol $\dfrac{dy}{dx}$, the part $\dfrac{d}{dx}$ taken separately represents the operator 'differentiate with respect to x'. This operator can be used in statements such as

$$\frac{d}{dx}(x^2) = 2x \quad \text{('The derivative with respect to } x \text{ of } x^2 \text{ is } 2x.\text{')}$$

or $\dfrac{d}{dx}(e^{3x}) = 3e^{3x}$

It can also be used to state, for example, the product rule: $\dfrac{d}{dx}(uv) = u\dfrac{dv}{dx} + v\dfrac{du}{dx}$

B1 Write each of these statements using the operator $\dfrac{d}{dx}$.

(a) The derivative with respect to x of $\ln x$ is $\dfrac{1}{x}$.

(b) The derivative with respect to x of $(x + \sin 3x)$ is $1 + 3\cos 3x$.

Return now to the equation $xy + y = 9$, whose graph is shown above.

The gradient function can be found by differentiating both sides of the equation with respect to x.

The first term xy is a product, so you need to use the product rule for this:

$$\frac{d}{dx}(xy) = x\frac{dy}{dx} + y\frac{dx}{dx} = x\frac{dy}{dx} + y$$

Differentiating the whole equation term by term with respect to x gives $xy + y = 9$.

$$x\frac{dy}{dx} + y + \frac{dy}{dx} = 0$$

This equation can be used to find the value of the gradient $\frac{dy}{dx}$ at a given point on the curve, by substituting the known values of x and y.

B2 The point $(2, 3)$ lies on the curve $xy + y = 9$.

    (a) By substituting $x = 2$, $y = 3$ into the derived equation $x\frac{dy}{dx} + y + \frac{dy}{dx} = 0$, find the value of $\frac{dy}{dx}$ at the point $(2, 3)$.

    (b) Show that for all values of x and y, $\frac{dy}{dx} = -\frac{y}{x+1}$.

The result in B2(b) can be confirmed by using a different method, as follows.

The equation $xy + y = 9$ can be rearranged as $y = \frac{9}{x+1}$, with y as an explicit function of x.

From this it follows that $\frac{dy}{dx} = -\frac{9}{(x+1)^2} = -\frac{9}{(x+1)(x+1)} = -\frac{y}{x+1}$.

However, it is not always possible to rewrite an implicit equation in explicit form. Even when it is possible, the result may be complicated. So you need to learn how to differentiate a function expressed implicitly.

This involves differentiating with respect to x expressions containing both x and y.

For example, $\frac{d}{dx}(x^2 + y) = \frac{d}{dx}(x^2) + \frac{d}{dx}(y) = 2x + \frac{dy}{dx}$

    or $\frac{d}{dx}(x^2y) = x^2\frac{d}{dx}(y) + y\frac{d}{dx}(x^2)$ (using the product rule)

$$= x^2\frac{dy}{dx} + y(2x) = x^2\frac{dy}{dx} + 2xy$$

B3 Write each of these in terms of x and $\frac{dy}{dx}$.

    (a) $\frac{d}{dx}(x^3 + y)$ **(b)** $\frac{d}{dx}(x^4 - y)$ **(c)** $\frac{d}{dx}\left(\frac{1}{x} + 2y\right)$

B4 Write each of these in terms of x, y and $\frac{dy}{dx}$.

    (a) $\frac{d}{dx}(x^3y)$ **(b)** $\frac{d}{dx}(x^4y)$ **(c)** $\frac{d}{dx}(e^xy)$

B5 The equation of a curve is $x^2y + x + y = 9$.

    (a) Show that $(x^2 + 1)\frac{dy}{dx} + 2xy + 1 = 0$.

    (b) Verify that the point $(1, 4)$ lies on the curve and find the gradient at this point.

The phrase 'differentiate y^2' is incomplete and therefore unclear.

It could mean 'differentiate y^2 *with respect to y*' or 'differentiate y^2 *with respect to x*'.

These are different. In symbols, the first is $\dfrac{d}{dy}(y^2)$ and the second is $\dfrac{d}{dx}(y^2)$.

The two operators $\dfrac{d}{dx}$ and $\dfrac{d}{dy}$ are connected by the chain rule:

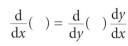

$$\frac{d}{dx}(\ ) = \frac{d}{dy}(\ )\frac{dy}{dx}$$

So, for example, $\dfrac{d}{dx}(y^2) = \dfrac{d}{dy}(y^2)\dfrac{dy}{dx} = 2y\dfrac{dy}{dx}$ 'To differentiate y^2 with respect to x, differentiate it with respect to y and then multiply by $\dfrac{dy}{dx}$.'

B6 Write each of these in terms of y and $\dfrac{dy}{dx}$.

(a) $\dfrac{d}{dx}(y^3)$ (b) $\dfrac{d}{dx}(y^4)$ (c) $\dfrac{d}{dx}(y-2)^3$ (d) $\dfrac{d}{dx}(2y+1)^2$ (e) $\dfrac{d}{dx}(y^{-2})$

Now we are in a position to differentiate with respect to x a product such as xy^2.

$$\frac{d}{dx}(xy^2) = x\frac{d}{dx}(y^2) + y^2\frac{d}{dx}(x) \quad \text{(using the product rule)}$$

$$= x\frac{d}{dy}(y^2)\frac{dy}{dx} + y^2(1) \quad \text{(using the chain rule for the first part)}$$

$$= x(2y)\frac{dy}{dx} + y^2$$

$$= 2xy\frac{dy}{dx} + y^2$$

B7 Write each of these in terms of x, y and $\dfrac{dy}{dx}$.

(a) $\dfrac{d}{dx}(xy^3)$ (b) $\dfrac{d}{dx}(x^2y^2)$ (c) $\dfrac{d}{dx}(x^3y^4)$ (d) $\dfrac{d}{dx}(x(y-1)^2)$

B8 A curve has the equation $y^2 + x + y = 7$.

(a) By differentiating both sides of this equation with respect to x, show that
$$(2y+1)\frac{dy}{dx} + 1 = 0$$

(b) Verify that the point $(1, 2)$ lies on the curve and find the value of $\dfrac{dy}{dx}$ at this point.

B9 A curve has the equation $xy^2 + y = 6$.

(a) By differentiating both sides of this equation with respect to x, show that
$$\frac{dy}{dx} = \frac{-y^2}{2xy+1}$$

(b) Verify that the point $(1, 2)$ lies on the curve and find the value of $\dfrac{dy}{dx}$ at this point.

Example 2

A curve is defined by the equation $2x^2 + xy + y^2 = 28$.
Find the equations of the tangent and the normal to the curve at the point $(3, 2)$
in the form $ax + by + c = 0$, where a, b and c are integers.

Solution

$$2x^2 + xy + y^2 = 28$$

Differentiate both sides of the equation
with respect to x.
$$\frac{d}{dx}\left(2x^2 + xy + y^2\right) = 0$$

Use the product rule for the term xy
and the chain rule for the term y².
$$4x + \left(x\frac{dy}{dx} + y\right) + 2y\frac{dy}{dx} = 0$$

Substituting $x = 3$ and $y = 2$,
$$12 + 3\frac{dy}{dx} + 2 + 4\frac{dy}{dx} = 0$$

$$\Rightarrow \qquad 14 + 7\frac{dy}{dx} = 0 \quad \text{so} \quad \frac{dy}{dx} = -2$$

The gradient of the tangent at $(3, 2)$ is -2.

So the equation of the tangent is $y - 2 = -2(x - 3)$, or $2x + y - 8 = 0$.

The gradient of the normal at $(3, 2)$ is $\frac{1}{2}$.

The equation of the normal is $y - 2 = \frac{1}{2}(x - 3)$, or $x - 2y + 1 = 0$.

Exercise B (answers p 167)

1 A curve is defined by the equation $x^2 + xy + y = 7$.

   (a) Show that $\dfrac{dy}{dx} = -\dfrac{2x + y}{x + 1}$

   (b) Find the equations of the tangent and the normal to the curve at the point $(2, 1)$.

2 Find, in the form $ax + by + c = 0$, the equation of **(i)** the tangent **(ii)** the normal
  to each of the curves defined below, at the given point.

   (a) $2x^2 - y^2 = 1$, at $(1, 1)$ **(b)** $x^3 + xy + y = 11$, at $(2, 1)$

   (c) $xy + x - y^2 = 9$, at $(5, 4)$ **(d)** $x^2y + xy^2 = 12$, at $(3, 1)$

   (e) $x^3 + y^2 = 9$, at $(2, 1)$ **(f)** $(x - 2)^2 + 3(y - 1)^2 = 16$, at $(4, 3)$

3 The curve whose equation is $x^2 + xy + (y - 4)^2 = 7$ crosses the line $x = 1$ at two points.

   (a) Find the coordinates of the two points.

   (b) Find the gradient of the curve at each of the two points.

4 The curve whose equation is $x^2 + xy + y^2 = 13$ crosses the line $y = 1$ at two points.

   (a) Find the coordinates of the two points.

   (b) Find the gradient of the curve at each of the two points.

   (c) Find the equation of the tangent at each of the two points.

Key points

- If x and y are defined in terms of a parameter t, then $\dfrac{dy}{dx} = \dfrac{\dfrac{dy}{dt}}{\dfrac{dx}{dt}}$. (p 47)

- In equations such as $xy + x + y^2 = 5$, $x^2 + 3y + xy = 6$ and so on, y is defined implicitly in terms of x. (p 50)

- The symbol $\dfrac{d}{dx}$ denotes the operator 'differentiate with respect to x'. (p 50)

- The chain rule is used when differentiating a function of y with respect to x:

$$\frac{d}{dx}(\ ) = \frac{d}{dy}(\ )\frac{dy}{dx}$$

  For example, $\dfrac{d}{dx}(y^2) = \dfrac{d}{dy}(y^2)\dfrac{dy}{dx} = 2y\dfrac{dy}{dx}$ (p 52)

- By differentiating both sides of the equation that defines a function implicitly, an equation for $\dfrac{dy}{dx}$ is obtained. (pp 52–53)

Mixed questions (answers p 167)

1 A curve is defined parametrically by the equations $x = 1 + 2t$, $y = 1 + \dfrac{1}{t}$ $(t > 0)$.

(a) (i) Find $\dfrac{dy}{dx}$ in terms of t.

  (ii) Find the value of $\dfrac{dy}{dx}$ when $t = 2$.

(b) A is the point on the curve for which $t = 2$. Find the coordinates of A.

(c) (i) Show that the relationship between x and y can be written in the form $xy - x - y = 1$.

  (ii) By differentiating both sides of this equation with respect to x, find the value of $\dfrac{dy}{dx}$ at the point A.

(d) Check that the answers obtained in (a) (ii) and (c) (ii) agree.

2 A curve C is defined parametrically by the equations $x = \cos\theta$, $y = \cos 2\theta$.

(a) Plot the points obtained by letting θ be 0, $\dfrac{\pi}{4}$, $\dfrac{\pi}{2}$, $\dfrac{3\pi}{4}$, ... up to 2π.

(b) What happens when θ takes values greater than 2π?

(c) Explain why the curve C consists of a part, but not the whole, of the graph of $y = 2x^2 - 1$.

(d) Find $\dfrac{dy}{dx}$ in terms of θ.

(e) Find the equation of the tangent to C at the point where $\theta = \dfrac{\pi}{4}$.

(f) Find the equation of the normal to C at the point where $\theta = \dfrac{\pi}{4}$.

3 The equation of the circle with centre $(2, 3)$ and radius 5 can be written as
$$(x-2)^2 + (y-3)^2 = 25$$

(a) Find an expression for $\dfrac{dy}{dx}$ as a function of x and y, simplifying your answer.

(b) Find the coordinates of the two points where this circle intersects the line $y = 2x - 6$.

(c) Find the equation of the tangent to the circle at each of these points.

(d) Find the equation of the normal to the circle at each of these points.

(e) Verify that each of the two normals passes through the centre of the circle.

4 A curve is defined by $\sqrt{x} + \sqrt{y} = 5 \ (x \geq 0, \ y \geq 0)$.

(a) Show that $\dfrac{dy}{dx} = -\sqrt{\dfrac{y}{x}}$.

(b) Hence find the equation of the tangent to the curve at the point where $x = 9$.

5 A curve is defined by parametric equations $x = \tan\theta + 1, \ y = \cos 4\theta \ \left(-\frac{1}{2}\pi \leq \theta < \frac{1}{2}\pi\right)$.

(a) Find the gradient of the curve at the point A where $\theta = \dfrac{\pi}{3}$.

(b) Find an equation of the normal to the curve at A.

(c) Find an equation of the normal to the curve at the point B where $\theta = \dfrac{\pi}{4}$.

6 A curve is defined by the equation $(x^2 - 1)y = x^2 + 1$.

(a) Show that $(x^2 - 1)\dfrac{dy}{dx} = 2x(1 - y)$.

(b) Hence find the gradient of the curve at the point where $x = 2$.

Test yourself (answers p 168)

1 A curve is defined by the equation $x^2 + 3xy + 2y^2 = 12$.
Find the equations of the tangent and the normal to the curve at the point $(2, 1)$ in the form $ax + by + c = 0$, where a, b and c are integers.

2 A curve is given by the parametric equations $x = \cos\theta, \ y = \ln(\sin\theta), \ 0 < \theta < \pi$.
Find the gradient of the tangent to the curve when $\theta = \frac{1}{4}\pi$.

3 A curve has equation $7x^2 + 48xy - 7y^2 + 75 = 0$.

A and B are two distinct points on the curve.
At each of these points the gradient of the curve is equal to $\frac{2}{11}$.

(a) Use implicit differentiation to show that $x + 2y = 0$ at the points A and B.

(b) Find the coordinates of the points A and B.

Edexcel

4 A curve is given by the parametric equations $x = 4\sin^3 t, \ y = \cos 2t, \ 0 \leq t \leq \dfrac{\pi}{4}$.

(a) Show that $\dfrac{dx}{dy} = -3\sin t$.

(b) Find an equation of the normal to the curve at the point where $t = \dfrac{\pi}{6}$.

Edexcel

5 Integration 1

In this chapter you will learn how to

- integrate e^x, $\frac{1}{x}$, $\sin x$ and $\cos x$
- use trigonometric identities to integrate functions such as $\sin^2 x$
- integrate by substitution and by parts

Key points from previous books

- Integration is the reverse of differentiation.

- $\int x^n \, dx = \dfrac{x^{n+1}}{n+1} + c \quad (n \neq -1)$

- The definite integral $\int_a^b f(x) \, dx$ gives the area under the
  graph of $y = f(x)$ between $x = a$ and $x = b$.

  Areas below the x-axis are negative.

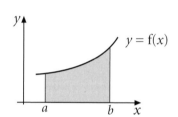

- $\cos^2 x + \sin^2 x = 1 \quad \sec^2 x = 1 + \tan^2 x \quad \operatorname{cosec}^2 x = 1 + \cot^2 x$
  $\sin 2x = 2 \sin x \cos x$
  $\cos 2x = \cos^2 x - \sin^2 x = 2\cos^2 x - 1 = 1 - 2\sin^2 x$

- If $\sin \theta = x$ and θ is in the range $-\frac{1}{2}\pi \leq \theta \leq \frac{1}{2}\pi$ then $\theta = \arcsin x$.
  If $\cos \theta = x$ and θ is in the range $0 \leq \theta \leq \pi$ then $\theta = \arccos x$.
  If $\tan \theta = x$ and θ is in the range $-\frac{1}{2}\pi < \theta < \frac{1}{2}\pi$ then $\theta = \arctan x$.

A Review

The questions that follow include some practice in trigonometric identities,
as these will be needed later in the chapter.

Exercise A (answers p 169)

1 Find the following indefinite integrals.

  (a) $\int (x^2 - x) \, dx$ 　　　　(b) $\int (5x^4 + 1) \, dx$ 　　　　(c) $\int \left(2x^3 - \frac{1}{2}x^2 - 3\right) dx$

2 Find $\int 3x^{-4} \, dx$.

3 Find the following indefinite integrals.

  (a) $\int \left(\dfrac{1}{x^2} + \dfrac{2}{x^3}\right) dx$ 　　　　(b) $\int \left(1 - \dfrac{1}{3x^2}\right) dx$ 　　　　(c) $\int \dfrac{6}{5x^3} \, dx$

4 Show that $\int_1^2 \left(\dfrac{3}{5}x^2 - \dfrac{1}{x^2}\right) dx = \dfrac{9}{10}$.

5 Find the following indefinite integrals.

(a) $\int 3x^{\frac{1}{2}}\,dx$ (b) $\int x^{\frac{4}{3}}\,dx$ (c) $\int \sqrt{x}\,dx$ (d) $\int \dfrac{1}{2\sqrt{x}}\,dx$ (e) $\int \sqrt{x^3}\,dx$

6 Evaluate the following definite integrals.

(a) $\int_1^3 (x^2 - 1)\,dx$ (b) $\int_{-1}^2 (2x^3 + 3)\,dx$ (c) $\int_{\frac{1}{2}}^2 \left(3x + \dfrac{1}{x^2}\right)dx$

(d) $\int_1^4 \dfrac{3}{\sqrt{x}}\,dx$ (e) $\int_0^1 5\sqrt{x}\,dx$ (f) $\int_0^8 \dfrac{1}{\sqrt[3]{x}}\,dx$

7 (a) Find $\int 3x(x-1)\,dx$.

   (b) Hence show that $\int_1^2 3x(x-1)\,dx = \frac{5}{2}$.

8 Find the following indefinite integrals.

(a) $\int 5x\left(x - \dfrac{1}{x}\right)dx$ (b) $\int (2x + 3)^2\,dx$ (c) $\int \left(x + \dfrac{1}{x}\right)^2 dx$

9 Evaluate $\int_1^2 \left(2x - \dfrac{1}{x^3}\right)^2 dx$, correct to four significant figures.

10 The diagram shows part of the graph of $y = (x+1)(3-x)$. Find the shaded area.

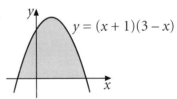

11 Find the following indefinite integrals.

(a) $\int \frac{1}{3}\sqrt{x}(\sqrt{x} + 2)\,dx$ (b) $\int \dfrac{x+1}{2x^3}\,dx$ (c) $\int \dfrac{x-1}{\sqrt{x}}\,dx$ (d) $\int \dfrac{x^2-3}{2\sqrt{x}}\,dx$

12 Prove each of the following identities.

(a) $\sin\theta(1 - \cos^2\theta) \equiv \sin^3\theta$ (b) $\dfrac{5 - 5\sin^2 x}{\cos^2 x} \equiv 5$

(c) $\cos^2 t(\sec^2 t - 1) \equiv \sin^2 t$ (d) $\sin 2\theta \tan\theta \equiv 2\sin^2\theta$

(e) $\dfrac{\sin 2\theta}{\sin\theta} \equiv 2\cos\theta$ (f) $\dfrac{\cos 2t}{\cos^2 t} \equiv 1 - \tan^2 t$

(g) $\dfrac{\cos 2\theta}{\sin^2\theta} \equiv \operatorname{cosec}^2\theta - 2$ (h) $\dfrac{6\sin\theta\cos\theta}{\cos 2\theta} \equiv 3\tan 2\theta$

13 The diagram shows part of the graph of $y = \frac{1}{4}x^2(4x - 9)$.

   (a) Evaluate $\int_0^3 \frac{1}{4}x^2(4x - 9)\,dx$.

   (b) What does the answer to part (a) tell you about the areas of regions A and B?

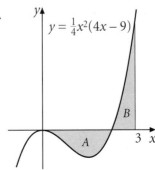

B Integrating e^x, sin x, cos x and $\sec^2 x$ (answers p 170)

The indefinite integrals of e^x, $\sin x$ and $\cos x$ can be found by 'reversing' the corresponding facts about their derivatives.

Because the derivative of e^x is e^x, the indefinite integral of e^x is $e^x + c$.

Because the derivative of $\sin x$ is $\cos x$, the indefinite integral of $\cos x$ is $\sin x + c$.

B1 (a) Write down the derivative of $\cos x$.

(b) Hence write down the indefinite integral of $\sin x$.

B2 Complete the calculation of each area below, giving exact values.

(a)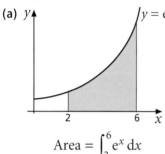

$$\text{Area} = \int_2^6 e^x \, dx$$

$$= \left[e^x \right]_2^6 = \dots$$

(b)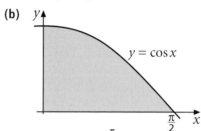

$$\text{Area} = \int_0^{\frac{\pi}{2}} \cos x \, dx$$

$$= \left[\dots \right]_0^{\frac{\pi}{2}} = \dots$$

The indefinite integrals of functions like $\sin 2x$, e^{3x}, and so on can be found by 'thinking backwards'.

B3 (a) Write down the derivative of $\sin 2x$.

(b) Hence find the indefinite integral of (i) $2\cos 2x$ (ii) $\cos 2x$

B4 (a) Write down the derivative of $\cos 5x$.

(b) Hence find the indefinite integral of (i) $-5\sin 5x$ (ii) $\sin 5x$

B5 (a) Write down the derivative of e^{2x}.

(b) Hence find the indefinite integral of (i) $2e^{2x}$ (ii) e^{2x}

The results obtained so far can be summarised as follows:

$$\int e^{ax} \, dx = \frac{1}{a} e^{ax} + c \qquad \int \cos ax \, dx = \frac{1}{a} \sin ax + c \qquad \int \sin ax \, dx = -\frac{1}{a} \cos ax + c$$

B6 Write down the indefinite integral of (a) e^{4x} (b) $\sin 6x$ (c) $\cos \frac{1}{2}x$

B7 (a) Differentiate $\tan 4x$. (b) Hence find $\int \sec^2 4x \, dx$.

$$\int \sec^2 ax \, dx = \frac{1}{a} \tan ax + c$$

B8 Find (a) $\int_0^{0.1} e^{5x} \, dx$ (b) $\int_0^{\frac{\pi}{4}} \cos 2x \, dx$ (c) $\int_{\frac{\pi}{4}}^{\frac{\pi}{2}} \sin 2x \, dx$ (d) $\int_0^{\frac{\pi}{2}} \sec^2 \frac{1}{2}x \, dx$

Example 1

Find the area under the graph of $y = 5 + 2\sin 3x$ between $x = 0$ and $x = \frac{1}{6}\pi$.

Solution

$$\text{Area} = \int_0^{\frac{1}{6}\pi} (5 + 2\sin 3x)\,dx = \left[5x + 2\left(-\tfrac{1}{3}\cos 3x\right)\right]_0^{\frac{1}{6}\pi}$$

$$= \left(\tfrac{5}{6}\pi - \tfrac{2}{3}\cos\tfrac{1}{2}\pi\right) - \left(5 \times 0 - \tfrac{2}{3}\cos 0\right)$$

$$= \left(\tfrac{5}{6}\pi - 0\right) - \left(0 - \tfrac{2}{3}\right) = \tfrac{5}{6}\pi + \tfrac{2}{3}$$

Exercise B (answers p 170)

1 (a) Write down the indefinite integral of e^{3x}.

(b) Find

  (i) $\displaystyle\int_0^1 e^{3x}\,dx$ (ii) $\displaystyle\int_0^2 \left(e^{3x} - 1\right)dx$ (iii) $\displaystyle\int_1^2 \left(e^{3x} - x\right)dx$

2 (a) Write down the indefinite integral of $\cos 4x$.

(b) Find

  (i) $\displaystyle\int_0^{\frac{1}{8}\pi} \cos 4x\,dx$ (ii) $\displaystyle\int_{-\frac{1}{8}\pi}^{\frac{1}{8}\pi} (1 - \cos 4x)\,dx$

3 Find the following indefinite integrals.

  (a) $\displaystyle\int (2\cos 3x - 5e^{2x})\,dx$ (b) $\displaystyle\int (4\sec^2 5x + 5\sin 4x)\,dx$ (c) $\displaystyle\int (e^{-x} + e^{-2x})\,dx$

4 Find the following indefinite integrals.

  (a) $\displaystyle\int (x^2 + e^{3x})\,dx$ (b) $\displaystyle\int (x - \sin 4x)\,dx$ (c) $\displaystyle\int \left(e^{-x} + \frac{1}{x^2}\right)dx$

  (d) $\displaystyle\int \left(\sin 5x - \sqrt{x}\right)dx$ (e) $\displaystyle\int (e^{2x} + \sec^2 2x)\,dx$ (f) $\displaystyle\int \left(\cos 3x - \frac{1}{x^3}\right)dx$

5 Find each of the following definite integrals.

  (a) $\displaystyle\int_0^2 \left(3x - e^{3x}\right)dx$ (b) $\displaystyle\int_0^\pi \left(x^2 + \sin\tfrac{1}{2}x\right)dx$ (c) $\displaystyle\int_{-\pi}^\pi \left(2 - 3\cos\tfrac{1}{2}x\right)dx$

  (d) $\displaystyle\int_0^4 \left(e^{-2x} + 2x\right)dx$ (e) $\displaystyle\int_0^4 \left(x^3 - e^{-x}\right)dx$ (f) $\displaystyle\int_0^{\frac{\pi}{2}} (2x - \sin 2x)\,dx$

6 Find each of the following definite integrals.

  (a) $\displaystyle\int_0^{\frac{1}{8}\pi} \sec^2 2x\,dx$ (b) $\displaystyle\int_0^\pi \left(\sin\tfrac{1}{2}x + \sec^2\tfrac{1}{4}x\right)dx$

7 Find the area under the graph of $y = x^2 - e^{-2x}$ between $x = 1$ and $x = 3$.

8 Find the area under the graph of $y = \sqrt{x} + e^{\frac{1}{2}x}$ between $x = 1$ and $x = 4$.

9 Show that if $\displaystyle\int_0^k e^{-\frac{1}{4}x}\,dx = 1$ then $k = 4\ln\tfrac{4}{3}$.

C Using trigonometric identities (answers p 170)

In Core 3 you met formulae for $\sin 2x$ and $\cos 2x$ in terms of $\sin x$ and $\cos x$:

$$\sin 2x = 2\sin x \cos x \quad (1) \qquad\qquad \cos 2x = \cos^2 x - \sin^2 x \quad (2)$$
$$= 2\cos^2 x - 1 \qquad (3)$$
$$= 1 - 2\sin^2 x \qquad (4)$$

These formulae enable us to express $\sin^2 x$, $\cos^2 x$ and $\sin x \cos x$ in terms of $\sin 2x$ or $\cos 2x$, thus making it possible to integrate them.

C1 (a) Use formula (1) above to express $\sin x \cos x$ in terms of $\sin 2x$.

(b) Hence find $\int \sin x \cos x \, dx$.

C2 (a) Use formula (3) above to express $\cos^2 x$ in terms of $\cos 2x$.

(b) Hence find $\int \cos^2 x \, dx$.

C3 By expressing $\sin^2 x$ in terms of $\cos 2x$, find $\int \sin^2 x \, dx$.

C4 (a) By replacing x by $\frac{1}{2}x$ in formula (4) above, show that $\sin^2 \frac{1}{2}x = \frac{1}{2}(1 - \cos x)$.

(b) Hence find

(i) $\int \sin^2 \frac{1}{2}x \, dx$ (ii) $\int_0^{\frac{\pi}{2}} \sin^2 \frac{1}{2}x \, dx$

Example 2

Find $\int_0^{\frac{\pi}{8}} \cos^2 2x \, dx$.

Solution

Use the 'double angle' formula $\cos 2A = 2\cos^2 A - 1$, with $A = 2x$.

$$\cos 4x = 2\cos^2 2x - 1$$

Rearranging, we get $\cos^2 2x = \frac{1}{2}(\cos 4x + 1)$.

So $\int_0^{\frac{\pi}{8}} \cos^2 2x \, dx = \int_0^{\frac{\pi}{8}} \frac{1}{2}(\cos 4x + 1) \, dx$

$$= \frac{1}{2}\left[\frac{1}{4}\sin 4x + x\right]_0^{\frac{\pi}{8}}$$

$$= \frac{1}{2}\left(\frac{1}{4}\sin\frac{\pi}{2} + \frac{\pi}{8}\right) - 0 = \frac{1}{8} + \frac{\pi}{16}$$

Exercise C (answers p 170)

1 Find

(a) $\int_0^{\pi} \sin^2 x \, dx$ (b) $\int_0^{\frac{\pi}{6}} \cos^2 x \, dx$ (c) $\int_0^{\frac{\pi}{3}} \cos^2 \frac{1}{2}x \, dx$

2 (a) Show that $(1 + \cos x)^2 = \frac{3}{2} + 2\cos x + \frac{1}{2}\cos 2x$.

(b) Hence find $\int (1 + \cos x)^2 \, dx$.

(c) Find $\int (1 + \sin x)^2 \, dx$.

3 (a) Show that $\int (\cos\theta + \sin\theta)^2 \, d\theta = \theta - \frac{1}{2}\cos 2\theta + c$.

(b) Find $\int_0^{\frac{\pi}{2}} (\cos\theta - \sin\theta)^2 \, d\theta$.

4 (a) Find $\int (\cos 2x + \sin 2x)^2 \, dx$ in its simplest form.

(b) Hence evaluate $\int_0^{\frac{\pi}{8}} (\cos 2x + \sin 2x)^2 \, dx$.

5 Show that $\int_{\frac{\pi}{4}}^{\frac{\pi}{2}} (\cos x(2 \sin x - 1)) \, dx = \frac{1}{2}(\sqrt{2} - 1)$.

6 (a) Use the identity $\sec^2 x = 1 + \tan^2 x$ to find $\int \tan^2 x \, dx$.

(b) Hence evaluate $\int_0^{\frac{\pi}{3}} 3 \tan^2 x \, dx$.

7 (a) Find $\int \tan^2 \frac{1}{2}x \, dx$.

(b) Show that $\int_0^{\frac{\pi}{2}} \tan^2 \frac{1}{2}x \, dx = 2 - \frac{\pi}{2}$.

8 Evaluate $\int_0^{\frac{\pi}{8}} \left(\tan^2 2x + 3\right) dx$.

9 (a) Find $\int \cos^2 3x \, dx$.

(b) Hence evaluate $\int_{-\frac{\pi}{6}}^{\frac{\pi}{6}} \cos^2 3x \, dx$.

10 Evaluate $\int_0^{\frac{\pi}{2}} 2 \sin^2 \frac{1}{4}x \, dx$.

D Integrating $\frac{1}{x}$ (answers p 171)

At first sight it would appear that because the derivative of $\ln x$ is $\frac{1}{x}$, then the indefinite integral of $\frac{1}{x}$ should be $\ln x + c$.

However, $\ln x$ is defined only for positive values of x and it looks as though there is no indefinite integral for $\frac{1}{x}$ when x is negative.

But although $\ln x$ is not defined for negative x, $\ln(-x)$ is defined, because $-x$ is positive when x is negative.

If $y = \ln(-x)$, then by the chain rule $\frac{dy}{dx} = -\frac{1}{-x} = \frac{1}{x}$.

So when x is negative, the indefinite integral of $\frac{1}{x}$ is $\ln(-x) + c$.

This can be confirmed by finding areas under the graph of $y = \frac{1}{x}$.

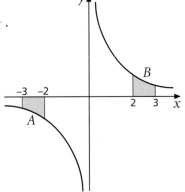

Area $A = \int_{-3}^{-2} \frac{1}{x} \, dx = \left[\ln(-x)\right]_{-3}^{-2} = \ln 2 - \ln 3$

Area $B = \int_2^3 \frac{1}{x} \, dx = \left[\ln x\right]_2^3 = \ln 3 - \ln 2$

As expected, $A = \ln 2 - \ln 3 = -(\ln 3 - \ln 2) = -B$.

The two statements $\int \frac{1}{x} dx = \ln x + c \quad (x > 0)$

$$\int \frac{1}{x} dx = \ln(-x) + c \quad (x < 0)$$

can be combined into a single statement using the modulus function:

 $\int \frac{1}{x} dx = \ln|x| + c \quad (x \neq 0)$

Notice the restriction $x \neq 0$. It means that it is impossible to find
the 'area' under $y = \frac{1}{x}$ between a negative and a positive value of x.

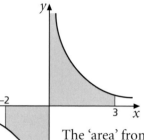

The 'area' from −2 to 3 is not defined as there is a discontinuity at $x = 0$.

D1 Find

(a) $\int_1^5 \frac{1}{x} dx$ (b) $\int_{-4}^{-2} \frac{1}{x} dx$ (c) $\int_2^5 \left(3 + \frac{1}{x}\right) dx$

D2 Find $\int \frac{1}{3x} dx$ by rewriting it as $\frac{1}{3}\int \frac{1}{x} dx$.

D3 (a) Differentiate $\ln(2x + 1)$ using the chain rule.

    (b) Hence show that $\int \frac{1}{2x + 1} dx = \frac{1}{2}\ln|2x + 1| + c$.

D4 (a) Differentiate $\ln(5x - 2)$ using the chain rule.

    (b) Hence find $\int \frac{1}{5x - 2} dx$.

Integrals of the form $\int \frac{1}{ax + b} dx$ can be found by 'thinking backwards' from $\ln(ax + b)$.

If you differentiate $\ln(ax + b)$, you get $\dfrac{a}{ax + b}$. It follows that

 $\int \frac{1}{ax + b} dx = \frac{1}{a}\ln|ax + b| + c$.

Example 3

Find $\int_2^5 \left(3 - \frac{2}{x}\right) dx$.

Solution

$$\int_2^5 \left(3 - \frac{2}{x}\right) dx = [3x - 2\ln|x|]_2^5$$

$$= (15 - 2\ln 5) - (6 - 2\ln 2)$$

$$= 9 - 2(\ln 5 - \ln 2)$$

$$= 9 - 2\ln\tfrac{5}{2}$$ *The last step uses the fact that* $\ln a - \ln b = \ln \dfrac{a}{b}$.

Example 4

Find $\displaystyle\int_1^3 \frac{x^2+1}{x}\,dx$.

Solution

First rewrite the expression as the sum of two separate fractions: $\displaystyle\frac{x^2+1}{x} = \frac{x^2}{x} + \frac{1}{x} = x + \frac{1}{x}$

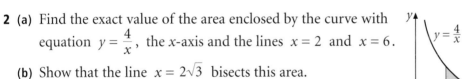

$$\int_1^3 \frac{x^2+1}{x}\,dx = \int_1^3 \left(x + \frac{1}{x}\right)dx = \left[\tfrac{1}{2}x^2 + \ln|x|\right]_1^3$$
$$= \left(\tfrac{9}{2} + \ln 3\right) - \left(\tfrac{1}{2} + \ln 1\right) = 4 + \ln 3$$

Exercise D (answers p 171)

1 Find the following indefinite integrals.

(a) $\displaystyle\int\left(1 + x - \frac{1}{x}\right)dx$

(b) $\displaystyle\int\left(e^{2x} + \frac{2}{x}\right)dx$

(c) $\displaystyle\int\left(\sin 6x - \frac{1}{2x}\right)dx$

(d) $\displaystyle\int\left(\frac{1}{x} + \sqrt{x}\right)dx$

(e) $\displaystyle\int\left(\frac{1}{x} + \frac{1}{\sqrt{x}}\right)dx$

(f) $\displaystyle\int\left(\cos 4x + \frac{1}{4x}\right)dx$

(g) $\displaystyle\int\left(\frac{5}{x} + e^{-3x}\right)dx$

(h) $\displaystyle\int\left(1 + \frac{1}{x} + \frac{1}{x^2} + \frac{1}{x^3}\right)dx$

(i) $\displaystyle\int\frac{1}{4x-1}\,dx$

2 (a) Find the exact value of the area enclosed by the curve with equation $y = \dfrac{4}{x}$, the x-axis and the lines $x = 2$ and $x = 6$.

(b) Show that the line $x = 2\sqrt{3}$ bisects this area.

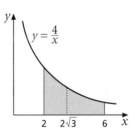

3 Find the exact value of each of these definite integrals.

(a) $\displaystyle\int_3^9 \frac{1}{2x}\,dx$

(b) $\displaystyle\int_1^4\left(\sqrt{x} - \frac{1}{x}\right)dx$

(c) $\displaystyle\int_1^2\left(e^{-3x} - \frac{1}{3x}\right)dx$

(d) $\displaystyle\int_2^8\left(x^2 + \frac{4}{x}\right)dx$

(e) $\displaystyle\int_4^9\left(\frac{1}{\sqrt{x}} - \frac{3}{2x}\right)dx$

(f) $\displaystyle\int_0^1\frac{1}{3x+2}\,dx$

4 (a) Write $\dfrac{1+x}{x^2}$ as the sum of two fractions.

(b) Hence find $\displaystyle\int_1^3\frac{1+x}{x^2}\,dx$.

5 Find $\displaystyle\int_1^3\frac{1+x^2}{x^3}\,dx$.

6 The curve $y = 6 - \dfrac{2}{x}$ crosses the x-axis at the point A.

(a) Find the x-coordinate of A.

(b) Find the shaded area bounded by the curve, the x-axis and the line $x = 1$.

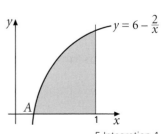

E Integration by substitution (answers p 172)

The integral $\int 2x(x^2 + 1)^5 \, dx$ could be done by multiplying out $2x(x^2 + 1)^5$, but this would be very time-consuming.

It is easier to introduce a new variable u such that $u = x^2 + 1$.

All the other parts of the integral, including dx, must now be rewritten in terms of u. We shall deal with dx first.

By differentiating $u = x^2 + 1$ we get $\dfrac{du}{dx} = 2x$.

Up to now we have treated the symbol $\dfrac{du}{dx}$ as a single symbol, not as a quantity du divided by a quantity dx.

But in the context of integration it is possible to write $\dfrac{du}{dx} = 2x$ as $du = 2x\,dx$. (The reason will appear later.)

The integral $\int 2x(x^2 + 1)^5 \, dx$ can be thought of as $\int (x^2 + 1)^5 \underbrace{2x\,dx}$.

Each part can now be expressed in terms of u.

This integral is easy to do.

Rewrite the result in terms of x.

$\int u^5 \, du$

$= \frac{1}{6}u^6 + c$

$= \frac{1}{6}(x^2 + 1)^6 + c$

The result can be checked by differentiation using the chain rule.

Let $y = \frac{1}{6}(x^2 + 1)^6 + c$.

Let $u = x^2 + 1$, so that $y = \frac{1}{6}u^6 + c$. Then $\dfrac{du}{dx} = 2x$ and $\dfrac{dy}{du} = u^5$.

So $\dfrac{dy}{dx} = \dfrac{dy}{du} \times \dfrac{du}{dx} = u^5 \times 2x = 2x(x^2 + 1)^5$.

The process of **integration by substitution** is based on the chain rule and this justifies rewriting $\dfrac{du}{dx} = 2x$ as $du = 2x\,dx$.

K

To integrate a function of x by substituting a new variable u:

(1) Find, by differentiation, the relationship between du and dx.

(2) Rewrite the function and dx in terms of of u and du.

(3) Carry out the integration in terms of u.

(4) Rewrite the result in terms of x.

E1 Follow the steps above to find $\int 3x^2(x^3 - 1) \, dx$ by using the substitution $u = x^3 - 1$.

E2 Use the substitution $u = 5 - x^2$ to find $\int 2x(5 - x^2)^3 \, dx$.

E3 Use the substitution $u = 5x + 1$ to find $\int \dfrac{5}{(5x + 1)^3} \, dx$.

Suppose we want to find $\int x(x^2-3)^4 \, dx$ using the substitution $u = x^2 - 3$.

Differentiating $u = x^2 - 3$ we get $\dfrac{du}{dx} = 2x$, from which $du = 2x \, dx$.

What appears in the integral is $x \, dx$, not $2x \, dx$.

However, this is easily dealt with: $\int x(x^2-3)^4 \, dx = \frac{1}{2}\int (x^2-3)^4 \, 2x \, dx = \frac{1}{2}\int u^4 \, du$

E4 Complete the process above to find $\int x(x^2-3)^4 \, dx$.

E5 Use the substitution $u = x^4 - 1$ to find $\int x^3(x^4-1)^3 \, dx$.

Sometimes the process of rewriting the function in terms of u is not so simple.

In this integral, let $u = 2x + 1$. $\qquad\qquad\qquad \int 2x(2x+1)^5 \, dx$

$\dfrac{du}{dx} = 2$, so $du = 2 \, dx$. $\qquad\qquad = \int x(2x+1)^5 \, 2 \, dx$

Replace $(2x+1)^5$ by u^5 and $2\,dx$ by du. $\quad = \int x u^5 \, du$
This still leaves x to be replaced.

$u = 2x + 1 \;\Rightarrow\; x = \dfrac{u-1}{2} \qquad\qquad = \int \left(\dfrac{u-1}{2}\right)u^5 \, du = \frac{1}{2}\int (u-1)u^5 \, du$

$$= \frac{1}{2}\int \left(u^6 - u^5\right) du$$

E6 Complete the process above to find $\int 2x(2x+1)^5 \, dx$.

E7 Use the substitution $u = 3x - 2$ to find $\int 6x(3x-2)^3 \, dx$.

E8 Use the substitution $u = 2x + 3$ to find $\int x(2x+3)^4 \, dx$.

Example 5
Use the substitution $u = 1 - x^2$ to find $\int x\sqrt{1-x^2} \, dx$.

Solution
First differentiate $u = 1 - x^2$. $\dfrac{du}{dx} = -2x$, so $du = -2x \, dx$.

Rearrange the integral so that $-2x \, dx$ appears in it.

$\displaystyle\int x\sqrt{1-x^2} \, dx = -\frac{1}{2}\int \sqrt{1-x^2}\,(-2x)\,dx$

$\displaystyle\qquad\qquad = -\frac{1}{2}\int \sqrt{u} \, du = -\frac{1}{2}\int u^{\frac{1}{2}} \, du$

$\displaystyle\qquad\qquad = -\frac{1}{2}\left(\frac{2}{3}u^{\frac{3}{2}}\right) + c = -\frac{1}{3}u^{\frac{3}{2}} + c$

Rewrite the result in terms of x. $\qquad = -\frac{1}{3}\left(1-x^2\right)^{\frac{3}{2}} + c$

Exercise E (answers p 172)

1 Use a suitable substitution to find $\int \dfrac{2}{(2x-7)^4}\,dx$.

2 Use the given substitution to find each of these indefinite integrals.

(a) $\int 3x^2(x^3-1)^4\,dx$ $\quad(u=x^3-1)$ (b) $\int x^2(x^3-1)^{-2}\,dx$ $\quad(u=x^3-1)$

(c) $\int x\sqrt{x^2-4}\,dx$ $\quad(u=x^2-4)$ (d) $\int x(2x^2+1)^5\,dx$ $\quad(u=2x^2+1)$

3 Use the given substitution to find each of these indefinite integrals.

(a) $\int 2x(2x-3)^4\,dx$ $\quad(u=2x-3)$ (b) $\int x(2x+1)^4\,dx$ $\quad(u=2x+1)$

(c) $\int x\sqrt{3x-2}\,dx$ $\quad(u=3x-2)$ (d) $\int \dfrac{5x}{5x-2}\,dx$ $\quad(u=5x-2)$

(e) $\int \dfrac{x}{4x+1}\,dx$ $\quad(u=4x+1)$ (f) $\int \dfrac{x}{(2x-1)^3}\,dx$ $\quad(u=2x-1)$

4 Find each of the following indefinite integrals. Choose your own substitution.

(a) $\int 2x(x^2+3)^3\,dx$ (b) $\int x\sqrt{x^2+5}\,dx$ (c) $\int x(6x+5)^5\,dx$

(d) $\int x(1-2x)^4\,dx$ (e) $\int \dfrac{x}{\sqrt{x^2-1}}\,dx$ (f) $\int \dfrac{x}{\sqrt{x-1}}\,dx$

F Further integration by substitution (answers p 172)

The diagram shows part of the graph of the function $y=x\sqrt{2x+1}$.

The area under the graph between $x=0$ and $x=4$ is given by

$$\int_0^4 x\sqrt{2x+1}\,dx$$

A suitable substitution is $u=2x+1$, from which $du=2dx$, so $dx=\tfrac{1}{2}du$.

As the integral is a definite integral, when you change from x to u, you change the lower and upper limits to values of u.

When $x=0$, $u=2\times0+1=1$.

When $x=4$, $u=2\times4+1=9$.

So the integral is rewritten as shown on the right.

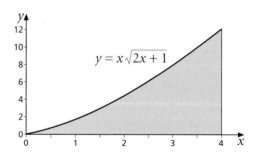

F1 Show that the integral above is equivalent to $\tfrac{1}{4}\int_1^9\left(u^{\frac{3}{2}}-u^{\frac{1}{2}}\right)du$ and find its value.

F2 Use the substitution $u = x + 1$ to find the value of $\int_0^3 \frac{x}{\sqrt{x+1}} \, dx$.

F3 Use the substitution $u = x^2 + 9$ to find the value of $\int_0^4 x\sqrt{x^2 + 9} \, dx$.

So far all substitutions have been expressed in the form $u = $ a function of x.
It is also possible to express a substitution in the form $x = $ a function of u.

The letter θ is often used when the substitution involves a trigonometric function.

For example, the integral $\int \frac{1}{\sqrt{1-x^2}} \, dx$ can be found by using the substitution $x = \sin\theta$,

where $\theta = \arcsin x$ and so θ is between $-\frac{\pi}{2}$ and $\frac{\pi}{2}$.

The expression $\sqrt{1-x^2}$ becomes $\sqrt{1 - \sin^2\theta} = \sqrt{\cos^2\theta} = \cos\theta$.

From $x = \sin\theta$ we get $\frac{dx}{d\theta} = \cos\theta$, so $dx = \cos\theta \, d\theta$.

So $\int \frac{1}{\sqrt{1-x^2}} \, dx = \int \frac{1}{\cos\theta} \cos\theta \, d\theta = \int 1 \, d\theta = \theta + c$.

So $\int \frac{1}{\sqrt{1-x^2}} \, dx = \arcsin x + c$.

In the case of a definite integral, there is no need to return to using x.
The lower and upper limits can be replaced by corresponding values of θ.

For example, in the integral $\int_0^1 \frac{1}{\sqrt{1-x^2}} \, dx$ the limits 0 and 1 are replaced by 0 and $\frac{1}{2}\pi$,

as these are the values of θ for which $\sin\theta = 0$ and $\sin\theta = 1$.

So $\int_0^1 \frac{1}{\sqrt{1-x^2}} \, dx = \int_0^{\frac{1}{2}\pi} \frac{1}{\cos\theta} \cos\theta \, d\theta = \int_0^{\frac{1}{2}\pi} 1 \, d\theta = [\theta]_0^{\frac{1}{2}\pi} = \frac{1}{2}\pi - 0 = \frac{1}{2}\pi$.

F4 Use the substitution $x = 3\sin\theta$ to find the value of $\int_0^3 \frac{1}{\sqrt{9-x^2}} \, dx$.

F5 Show that the substitution $x = \tan\theta$ transforms the integral $\int_0^1 \frac{1}{1+x^2} \, dx$
into $\int_0^{\frac{1}{4}\pi} 1 \, d\theta$ and hence find its value.

F6 Show that the substitution $x = \tan\theta$ transforms the integral $\int_0^1 \frac{1}{\left(1+x^2\right)^2} \, dx$
into $\int_0^{\frac{1}{4}\pi} \cos^2\theta \, d\theta$ and hence find its value.

F7 (a) By using the substitution $x = \sin\theta$, show that $\int_0^1 x^2\sqrt{1-x^2} \, dx = \int_0^{\frac{1}{2}\pi} \sin^2\theta\cos^2\theta \, d\theta$.

(b) Show that $\sin^2\theta\cos^2\theta = \frac{1}{4}\sin^2 2\theta$.

(c) Hence evaluate $\int_0^{\frac{1}{2}\pi} \sin^2\theta\cos^2\theta \, d\theta$.

Example 6

Use the substitution $x = \sin\theta$ to evaluate $\int_0^1 \sqrt{1-x^2}\, dx$.

Solution

$\sqrt{1-x^2} = \sqrt{1-\sin^2\theta} = \cos\theta$, $dx = \cos\theta\, d\theta$. When $x = 0$, $\theta = 0$. When $x = 1$, $\theta = \tfrac{1}{2}\pi$.

$$\int_0^1 \sqrt{1-x^2}\, dx = \int_0^{\frac{1}{2}\pi} \cos\theta\cos\theta\, d\theta = \int_0^{\frac{1}{2}\pi} \cos^2\theta\, d\theta = \int_0^{\frac{1}{2}\pi} \tfrac{1}{2}(\cos 2\theta + 1)\, d\theta$$

$$= \tfrac{1}{2}\left[\tfrac{1}{2}\sin 2\theta + \theta\right]_0^{\frac{1}{2}\pi} = \tfrac{1}{2}\left(\tfrac{1}{2}\sin\pi + \tfrac{1}{2}\pi\right) = \tfrac{1}{4}\pi$$

Exercise F (answers p 173)

1 Use the substitutions given in brackets to evaluate the following.

(a) $\int_1^2 2x(3x+1)^3\, dx$ $(u = 3x+1)$ (b) $\int_{\sqrt{3}}^2 x\sqrt{x^2-3}\, dx$ $\left(u = x^2 - 3\right)$

(c) $\int_3^7 x\sqrt{x-3}\, dx$ $(u = x-3)$ (d) $\int_0^1 xe^{x^2}\, dx$ $\left(u = x^2\right)$

(e) $\int_0^2 \dfrac{1}{\sqrt{4-x^2}}\, dx$ $(x = 2\sin\theta)$ (f) $\int_0^{\frac{\pi}{2}} \cos^3 x\, dx$ $(u = \sin x)$

(g) $\int_{\frac{1}{2}}^{\frac{\sqrt{3}}{2}} \dfrac{\sqrt{1-x^2}}{x^2}\, dx$ $(x = \cos\theta)$ (h) $\int_{-\frac{\pi}{4}}^{\frac{\pi}{4}} \sec^4 x\, dx$ $(u = \tan x)$

2 Use the substitution $u = 1 + \sin\theta$ to show that $\int_0^{\frac{\pi}{2}} \dfrac{\sin 2\theta}{1 + \sin\theta}\, d\theta = 2(1 - \ln 2)$.

G Integrating $\dfrac{f'(x)}{f(x)}$ (answers p 173)

G1 Use the substitution $u = x^2 + 1$ to find $\int \dfrac{2x}{x^2+1}\, dx$.

G2 Use the substitution $u = x^3 - 1$ to find $\int \dfrac{3x^2}{x^3-1}\, dx$.

G3 Use the substitution $u = x^4 + 3x^2 - 5$ to find $\int \dfrac{4x^3 + 6x}{x^4 + 3x^2 - 5}\, dx$.

In all three of the previous questions, the integral, in terms of u, comes out as $\ln|u| + c$. This happens because, in the function being integrated, the numerator is the derivative of the denominator.

In symbols, each integral above is of the form $\int \dfrac{f'(x)}{f(x)}\, dx$.

When you use the substitution $u = f(x)$, then $\dfrac{du}{dx} = f'(x)$ and so $du = f'(x)\, dx$.

So the integral becomes $\int \dfrac{1}{u}\, du$, which is $\ln|u| + c$ or, in terms of x, $\ln|f(x)| + c$.

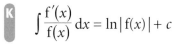

$\int \dfrac{f'(x)}{f(x)}\, dx = \ln|f(x)| + c$

Sometimes it is not immediately obvious that a function to be integrated can be expressed in the form $\dfrac{f'(x)}{f(x)}$. A little ingenuity may be needed, as in the next example.

Example 7

Find $\displaystyle\int \dfrac{x^3}{x^4+1}\,dx$.

Solution

The derivative of the denominator, $x^4 + 1$, is $4x^3$, so think of the numerator as $\tfrac{1}{4}(4x^3)$.

$$\int \dfrac{x^3}{x^4+1}\,dx = \tfrac{1}{4}\int \dfrac{4x^3}{x^4+1}\,dx = \tfrac{1}{4}\ln|x^4+1| + c$$

Exercise G (answers p 173)

1 Find each of these indefinite integrals.

(a) $\displaystyle\int \dfrac{3x^2}{x^3-1}\,dx$

(b) $\displaystyle\int \dfrac{2x+1}{x^2+x}\,dx$

(c) $\displaystyle\int \dfrac{x^4}{x^5+1}\,dx$

(d) $\displaystyle\int \dfrac{x-1}{x^2-2x}\,dx$

(e) $\displaystyle\int \dfrac{x+1}{2x^2+4x+1}\,dx$

(f) $\displaystyle\int \dfrac{e^x}{3e^x-1}\,dx$

2 (a) Find the indefinite integral $\displaystyle\int \dfrac{x}{4-x^2}\,dx$.

(b) Hence show that $\displaystyle\int_0^1 \dfrac{x}{4-x^2}\,dx = \tfrac{1}{2}\ln\tfrac{4}{3}$.

3 (a) Show that the area under the graph of $y = \dfrac{x}{x^2+1}$ between $x = 0$ and $x = 2$ is $\tfrac{1}{2}\ln 5$.

(b) Find the area under the graph of $y = \dfrac{x^2}{x^3+1}$ between $x = 0$ and $x = 2$.

4 (a) Express $\tan x$ in terms of $\sin x$ and $\cos x$.

(b) Explain why $\displaystyle\int \tan x\,dx = -\ln|\cos x| + c$.

(c) Find $\displaystyle\int \cot x\,dx$.

5 (a) Find $\displaystyle\int \dfrac{e^{3x}}{e^{3x}+1}\,dx$.

(b) Hence show that $\displaystyle\int_0^1 \dfrac{e^{3x}}{e^{3x}+1}\,dx = \tfrac{1}{3}\ln\left(\dfrac{e^3+1}{2}\right)$.

6 (a) Write down the derivative of $1 + \sqrt{x}$.

(b) Find $\displaystyle\int \dfrac{1}{\sqrt{x}\left(1+\sqrt{x}\right)}\,dx$.

7 Show that $\displaystyle\int_0^{\frac{\pi}{2}} \dfrac{\cos\theta}{3+\sin\theta}\,d\theta = \ln\tfrac{4}{3}$.

H Integration by parts (answers p 174)

Integration by parts is the name given to a method of integration that is based on using the product rule for differentiation 'backwards'.
An example will make this clear.

Suppose we start with a function that is a product of two functions, for example $x \sin x$.

The derivative of $x \sin x$ can be found by the product rule by letting $u = x$ and $v = \sin x$.

So the derivative of $x \sin x$ equals $u \dfrac{dv}{dx} + v \dfrac{du}{dx} = x \cos x + \sin x$.

Because integration is the reverse of differentiation, it follows that

$$\int (x \cos x + \sin x)\, dx = x \sin x + c$$

Rewriting the integral in two parts, we get

$$\int x \cos x\, dx + \int \sin x\, dx = x \sin x + c$$

The second of these integrals we know already, so subtract it from both sides to get

$$\int x \cos x\, dx = x \sin x - \int \sin x\, dx + c$$
$$= x \sin x + \cos x + c$$

By starting with the product $x \sin x$ we have found the integral of $x \cos x$.

H1 (a) By using the product rule to find the derivative of $x \cos x$, show that
$$\int (-x \sin x + \cos x)\, dx = x \cos x + c.$$
(b) Hence show that $\int x \sin x\, dx = \sin x - x \cos x + c.$

H2 (a) By using the product rule to find the derivative of $x e^x$, show that
$$\int (x e^x + e^x)\, dx = x e^x + c.$$
(b) Hence show that $\int x e^x\, dx = x e^x - e^x + c.$

H3 (a) By differentiating $x \sin 2x$, show that
$$\int (2x \cos 2x + \sin 2x)\, dx = x \sin 2x + c.$$
(b) Hence find $\int 2x \cos 2x\, dx.$

In the examples looked at so far, we started by differentiating a product and this led us to be able to find an integral.

However, in practice we start from an integral that we want to find and we don't know which product will lead to it. We shall now see how the product idea can be used to find a given integral.

The general form of the product rule is: derivative of $uv = u\dfrac{dv}{dx} + v\dfrac{du}{dx}$

Rewriting this in terms of integrals we get: integral of $\left(u\dfrac{dv}{dx} + v\dfrac{du}{dx}\right) = uv + c$.

This can also be written as $\displaystyle\int\left(u\dfrac{dv}{dx} + v\dfrac{du}{dx}\right)dx = uv + c$

or $\displaystyle\int u\dfrac{dv}{dx}\,dx + \int v\dfrac{du}{dx}\,dx = uv + c$

Now suppose the first integral is the one we want to find and the second is one that we already know how to find. By subtracting the known one from both sides, we get the **formula for integration by parts**:

$$\int u\dfrac{dv}{dx}\,dx = uv - \int v\dfrac{du}{dx}\,dx + c$$

An example will show how the formula is used.

Suppose we want to find $\displaystyle\int x\cos 3x\,dx$.

Compare this with $\displaystyle\int u\dfrac{dv}{dx}\,dx$.

Let $u = x$ and $\dfrac{dv}{dx} = \cos 3x$.

Then $v = \frac{1}{3}\sin 3x$ ('+ c' is not needed here.)

For the right-hand side of the formula we need $\dfrac{du}{dx}$.

Since $u = x$, $\dfrac{du}{dx} = 1$.

So applying the formula we get $\displaystyle\int u\dfrac{dv}{dx}\,dx = uv - \int v\dfrac{du}{dx}\,dx + c$

$= x\left(\tfrac{1}{3}\sin 3x\right) - \int\left(\tfrac{1}{3}\sin 3x\right)1\,dx + c$

$= \tfrac{1}{3}x\sin 3x + \tfrac{1}{9}\cos 3x + c$

H4 Find $\displaystyle\int x\sin 4x\,dx$ by integration by parts, letting $u = x$ and $\dfrac{dv}{dx} = \sin 4x$.

H5 Find $\displaystyle\int xe^{4x}\,dx$ by integration by parts, letting $u = x$ and $\dfrac{dv}{dx} = e^{4x}$.

H6 Find (a) $\displaystyle\int x\cos 4x\,dx$ (b) $\displaystyle\int xe^{5x}\,dx$ (c) $\displaystyle\int xe^{-4x}\,dx$

H7 What happens if you try to find $\displaystyle\int x\sin x\,dx$ by letting $u = \sin x$ and $\dfrac{dv}{dx} = x$?

H8 (a) If you try to find $\displaystyle\int x\ln x\,dx$ by letting $u = x$ and $\dfrac{dv}{dx} = \ln x$, you come up against a difficulty. What is it?

(b) Try letting $u = \ln x$ and $\dfrac{dv}{dx} = x$.

When you are finding a definite integral using integration by parts, you can deal with the upper and lower limits of each part separately:

 $\displaystyle\int_a^b u\frac{dv}{dx}\,dx = \left[uv\right]_a^b - \int_a^b v\frac{du}{dx}\,dx$

Example 8

Find $\displaystyle\int_0^2 xe^{3x}\,dx$.

Solution

Let $u = x$ and $\dfrac{dv}{dx} = e^{3x}$. So $v = \int e^{3x}\,dx = \frac{1}{3}e^{3x}$ and $\dfrac{du}{dx} = 1$.

$$\int_a^b u\frac{dv}{dx}\,dx = \left[uv\right]_a^b - \int_a^b v\frac{du}{dx}\,dx$$

$$\int_0^2 xe^{3x}\,dx = \left[x\left(\tfrac{1}{3}e^{3x}\right)\right]_0^2 - \int_0^2 \left(\tfrac{1}{3}e^{3x}\right)1\,dx$$

$$= \left(\tfrac{2}{3}e^6 - 0\right) - \tfrac{1}{3}\int_0^2 e^{3x}\,dx$$

$$= \tfrac{2}{3}e^6 - \tfrac{1}{3}\left[\tfrac{1}{3}e^{3x}\right]_0^2$$

$$= \tfrac{2}{3}e^6 - \tfrac{1}{9}\left(e^6 - 1\right) = \tfrac{1}{9}\left(6e^6 - e^6 + 1\right) = \tfrac{1}{9}\left(5e^6 + 1\right)$$

Exercise H (answers p 174)

1 Find each of these indefinite integrals.

(a) $\displaystyle\int 2x\cos 7x\,dx$ 　　(b) $\displaystyle\int 3xe^{6x}\,dx$ 　　(c) $\displaystyle\int x\sin\tfrac{1}{2}x\,dx$

(d) $\displaystyle\int xe^{-x}\,dx$ 　　(e) $\displaystyle\int xe^{-3x}\,dx$ 　　(f) $\displaystyle\int xe^{-\frac{1}{2}x}\,dx$

2 Find each of these definite integrals.

(a) $\displaystyle\int_0^{\frac{\pi}{2}} x\cos x\,dx$ 　　(b) $\displaystyle\int_0^{\pi} x\sin x\,dx$ 　　(c) $\displaystyle\int_0^4 xe^x\,dx$

(d) $\displaystyle\int_0^{\frac{\pi}{6}} x\sin 3x\,dx$ 　　(e) $\displaystyle\int_0^1 2xe^{-x}\,dx$ 　　(f) $\displaystyle\int_0^6 xe^{\frac{1}{2}x}\,dx$

3 This diagram shows the graph of $y = x\sin 2x$.

(a) What are the exact values of the x-coordinates of the three points in the diagram where the curve intersects the x-axis?

(b) Find the area of each of the regions labelled A and B.

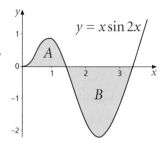

4 Evaluate $\displaystyle\int_1^3 x^3\ln x\,dx$.

5 (a) Use integration by parts to find $\displaystyle\int \ln x\,dx$ (let $u = \ln x$ and $\dfrac{dv}{dx} = 1$).

(b) Hence find $\displaystyle\int_1^5 \ln x\,dx$.

6 Evaluate $\displaystyle\int_1^3 \ln(2x)\,dx$ correct to three significant figures.

I Further integration by parts (answers p 174)

Sometimes it is necessary to integrate by parts twice.

Take, for example, the indefinite integral $\int x^2 e^x \, dx$.

Let $u = x^2$ and $\dfrac{dv}{dx} = e^x$, from which $v = e^x$. So

$$\int u \frac{dv}{dx} \, dx = uv - \int v \frac{du}{dx} \, dx$$

$$\int x^2 e^x \, dx = x^2 e^x - \int e^x (2x) \, dx$$

$$= x^2 e^x - 2\int x e^x \, dx$$

The integral $\int x e^x \, dx$ that appears on the right-hand side can itself be found by integration by parts.

I1 **(a)** Use integration by parts to find $\int x e^x \, dx$.

(b) Hence show that $\int x^2 e^x \, dx = e^x(x^2 - 2x + 2) + c$.

I2 **(a)** Show that $\int x^2 \cos x \, dx = x^2 \sin x - 2\int x \sin x \, dx$.

(b) Use integration by parts to find $\int x \sin x \, dx$.

(c) Hence find $\int x^2 \cos x \, dx$.

I3 Find $\int x^2 e^{4x} \, dx$.

I4 **(a)** Show that $\int_0^{\frac{\pi}{4}} x^2 \sin 2x \, dx = \left[-\tfrac{1}{2}x^2 \cos 2x\right]_0^{\frac{\pi}{4}} + \int_0^{\frac{\pi}{4}} x \cos 2x \, dx$.

(b) Show that $\int_0^{\frac{\pi}{4}} x \cos 2x \, dx = \dfrac{\pi}{8} - \tfrac{1}{4}$.

(c) Hence find the value of $\int_0^{\frac{\pi}{4}} x^2 \sin 2x \, dx$.

Example 9

Find $\int x^2 e^{-2x} \, dx$.

Solution

Let $u = x^2$ and $\dfrac{dv}{dx} = e^{-2x}$. So $v = -\tfrac{1}{2}e^{-2x}$.

$$\int u \frac{dv}{dx} \, dx = uv - \int v \frac{du}{dx} \, dx$$

$$\int x^2 e^{-2x} \, dx = -\tfrac{1}{2}x^2 e^{-2x} - \int (-\tfrac{1}{2}e^{-2x})(2x) \, dx$$

$$= -\tfrac{1}{2}x^2 e^{-2x} + \int x e^{-2x} \, dx$$

To find $\int x e^{-2x} \, dx$, let $u = x$ and $\dfrac{dv}{dx} = e^{-2x}$.

$$\int x e^{-2x} \, dx = -\tfrac{1}{2}x e^{-2x} - \int (-\tfrac{1}{2}e^{-2x}) \, dx$$

$$= -\tfrac{1}{2}x e^{-2x} + \tfrac{1}{2}\int e^{-2x} \, dx$$

$$= -\tfrac{1}{2}x e^{-2x} - \tfrac{1}{4}e^{-2x} + c$$

So $\int x^2 e^{-2x} \, dx = -\tfrac{1}{2}x^2 e^{-2x} - \tfrac{1}{2}x e^{-2x} - \tfrac{1}{4}e^{-2x} + c$

Exercise I (answers p 175)

1 Find these indefinite integrals.

(a) $\int 2x^2 \sin x \, dx$ (b) $\int \frac{1}{2}x^2 \cos 2x \, dx$ (c) $\int x^2 e^{-x} \, dx$

2 (a) Find $\int x^2 e^{2x} \, dx$.

(b) Hence show that $\int_0^2 x^2 e^{2x} \, dx = \frac{1}{4}(5e^4 - 1)$.

3 Evaluate $\int_0^\pi \frac{1}{2}x^2 \cos \frac{1}{2}x \, dx$.

4 The diagram shows part of the graph of $y = 3x^2 \sin 3x$.
Find the area of the shaded region.

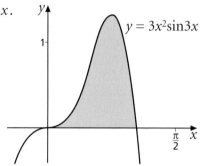

$y = 3x^2\sin3x$

Key points

- $\int e^{ax} \, dx = \frac{1}{a} e^{ax} + c$ $\int \cos ax \, dx = \frac{1}{a} \sin ax + c$ $\int \sin ax \, dx = -\frac{1}{a} \cos ax + c$

  $\int \sec^2 ax \, dx = \frac{1}{a} \tan ax + c$ (p 58)

- $\int \frac{1}{x} \, dx = \ln|x| + c$ $(x \neq 0)$ $\int \frac{1}{ax+b} \, dx = \frac{1}{a} \ln|ax + b| + c$ (p 62)

- To integrate a function of x by substituting a new variable u:
  (1) Find, by differentiation, the relationship between du and dx.
  (2) Rewrite the function and dx in terms of u and du.
  (3) Carry out the integration in terms of u.
  (4) Rewrite the result in terms of x.   (p 64)

- $\int \frac{f'(x)}{f(x)} \, dx = \ln|f(x)| + c$ (p 68)

- $\int u \frac{dv}{dx} \, dx = uv - \int v \frac{du}{dx} \, dx + c$ $\int_a^b u \frac{dv}{dx} \, dx = [uv]_a^b - \int_a^b v \frac{du}{dx} \, dx$ (integration by parts)

    (pp 70–71)

The following table should help you to decide which method of integration to use.

Description	Examples	Method
The function is a sum of individual functions each of which you know how to integrate.	$\int\left(x^2 + \dfrac{2}{x} - e^{4x} + \cos 5x\right)dx$	Integrate term by term.
The function is of the form $\dfrac{f'(x)}{f(x)}$ or can be rearranged into this form.	$\int\dfrac{x^2}{x^3-1}\,dx = \tfrac{1}{3}\int\dfrac{3x^2}{x^3-1}\,dx$	Identify f(x) to obtain $\ln\lvert f(x)\rvert + c$.
The function is a product of the form xe^{ax}, $x\cos ax$, $x\sin ax$, $x\ln x$ (or any of a similar type involving linear expressions, such as $(2x+3)e^{-4x}$, $(x+1)\cos 2x$, $x\sin(3x-1)$).	$\int x\sin 3x\,dx$ $\int xe^{-2x}\,dx$	Use integration by parts.
The function is a product of the form x^2e^{ax}, $x^2\cos ax$ or $x^2\sin ax$.	$\int x^2 e^{4x}\,dx$ $\int x^2\cos x\,dx$	Use integration by parts twice.
The function includes a function of a function, such as $(x^2+3)^5$, $\sqrt{2x-1}$, ...	$\int x\left(x^2+3\right)^5 dx$ $\int\dfrac{x}{\sqrt{2x+1}}\,dx$	Try substitution. (For the examples use $u = x^2 + 3$ $u = 2x + 1$)

Mixed questions (answers p 175)

1 A is the point on the graph of $y = \sin 2x$ where $x = \tfrac{5}{12}\pi$.

(a) Explain why the area of the triangle labelled T is $\tfrac{5}{48}\pi$.

(b) Hence find the shaded area labelled S.

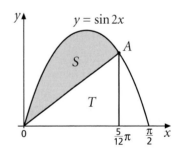

2 (a) Draw a rough sketch of the curve with equation $y = 2e^{\frac{1}{2}x}$ for $x \geq 0$.

(b) The curve crosses the line $y = 6$ at the point A.
Show that the x-coordinate of A is $2\ln 3$.

(c) Find the exact value of the area enclosed by the curve, the y-axis and the line $y = 6$.

3 Find the following indefinite integrals.

(a) $\int\sin(3x-1)\,dx$

(b) $\int\dfrac{1}{6x-1}\,dx$

(c) $\int\sec^2\tfrac{1}{2}x\,dx$

4 Find

(a) $\int\dfrac{3}{(4+3x)^5}\,dx$

(b) $\int\dfrac{x}{x^2-5}\,dx$

(c) $\int\dfrac{2}{(x-5)^2}\,dx$

5 Find the following definite integrals.

(a) $\displaystyle\int_0^\pi \sin\tfrac{1}{3}x\,dx$

(b) $\displaystyle\int_0^\pi \left(x + \sin\tfrac{1}{2}x\right)dx$

(c) $\displaystyle\int_4^9 \frac{x^{\frac{1}{2}}}{x^{\frac{3}{2}} - 1}\,dx$

6 The curve $y = e^x$ and the line $y = 2x + 1$ both pass through $(0, 1)$ and intersect again at the point where $x = a$.

(a) Write down the equation satisfied by a.

(b) Show that the area enclosed between the line and the curve is $a^2 - a$.

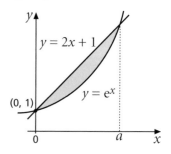

7 (a) Show that $\cos^2\tfrac{1}{2}x = \tfrac{1}{2}(1 + \cos x)$.

(b) Hence evaluate $\displaystyle\int_0^\pi \cos^2\tfrac{1}{2}x\,dx$.

8 Show that $\displaystyle\int_0^{\frac{\pi}{3}} \tan^2 x\,dx = \sqrt{3} - \frac{\pi}{3}$.

9 Use integration by parts to find these indefinite integrals.

(a) $\displaystyle\int x\sin 4x\,dx$

(b) $\displaystyle\int x^4 \ln x\,dx$

(c) $\displaystyle\int x^2 e^{3x}\,dx$

10 Show that $\displaystyle\int_1^2 \frac{x+1}{x^2 + 2x + 1}\,dx = \ln\tfrac{3}{2}$.

11 (a) Find the x-coordinate of the point where the curve $y = 2 + \dfrac{1}{x-2}$ crosses the x-axis.

(b) Find the area enclosed by the curve and the axes.

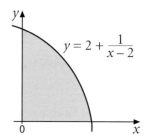

12 Evaluate $\displaystyle\int_{\frac{\pi}{4}}^{\frac{\pi}{2}} \frac{\sin 2x}{1 - \cos 2x}\,dx$.

13 Use the substitution $u = x^3$ to find $\displaystyle\int x^2 e^{x^3}\,dx$.

14 Use integration by parts to evaluate $\displaystyle\int_1^2 2x^5 \ln x\,dx$.

15 Use the substitution $x = \tfrac{2}{3}\tan\theta$ to find $\displaystyle\int \frac{1}{4 + 9x^2}\,dx$.

16 Use the substitution $x = \sin t$ to evaluate $\displaystyle\int_0^{\frac{\sqrt{3}}{2}} \frac{x^2}{\sqrt{1 - x^2}}\,dx$.

17 Find the following indefinite integrals.

(a) $\displaystyle\int xe^{-2x}\,dx$

(b) $\displaystyle\int x(x-2)^4\,dx$

(c) $\displaystyle\int \frac{3\sin\theta}{1 - \cos\theta}\,d\theta$

(d) $\displaystyle\int \frac{\ln x}{x^2}\,dx$

(e) $\displaystyle\int 4xe^{(x^2 + 5)}$

(f) $\displaystyle\int x^2 e^{\frac{1}{5}x}\,dx$

Test yourself (answers p 176)

1 Find the following indefinite integrals.

(a) $\int e^{6x}\,dx$ (b) $\int \cos\frac{1}{4}x\,dx$ (c) $\int \sin(2x+1)\,dx$

2 Find the following definite integrals.

(a) $\int_1^3 \frac{6}{x}\,dx$ (b) $\int_0^{\frac{1}{2}} e^{-2x}\,dx$ (c) $\int_0^4 \frac{1}{3x+2}\,dx$

(d) $\int_1^e \left(x-\frac{1}{x}\right)dx$ (e) $\int_0^{\frac{\pi}{4}} (\cos 2x + \sin 4x)\,dx$

3 Find the value of $\int_1^2 \frac{2x+1}{x^2+x+1}\,dx$.

4 Find the following indefinite integrals.

(a) $\int (\tan^2\theta + 5)\,d\theta$ (b) $\int \sin x \cos x\,dx$ (c) $\int \frac{3}{\cos^2 t}\,dt$

5 Use integration by parts to find $\int (2x+1)\sin 4x\,dx$.

6 Use the substitution $u = x+1$ to find $\int \frac{x^2-1}{\sqrt{x+1}}\,dx$, giving your answer in terms of x.

7 Use integration by parts to find the value of $\int_0^2 xe^{\frac{1}{2}x}\,dx$.

8 Use the substitution $u = 2 - \sin x$ to prove that

$$\int \frac{\sin 2x}{(2-\sin x)^3}\,dx = \frac{2(\sin x - 1)}{(2-\sin x)^2} + c,\ \text{where } c \text{ is an arbitrary constant.}$$

9 (a) Use integration by parts to find the exact value of $\int_1^3 x^2 \ln x\,dx$.

(b) Use the substitution $x = \sin\theta$ to show that, for $|x| \le 1$,

$$\int \frac{1}{(1-x^2)^{\frac{3}{2}}}\,dx = \frac{x}{(1-x^2)^{\frac{1}{2}}} + c,\ \text{where } c \text{ is an arbitrary constant.}$$ Edexcel

10 Use integration by parts to find $\int 2x^2 \sin\frac{1}{4}x\,dx$.

11 Use the substitution $x = 2\sqrt{3}\sin\theta$ to prove that $\int_{\sqrt{6}}^{2\sqrt{3}} \sqrt{12-x^2}\,dx = 3\left(\frac{\pi}{2}-1\right)$.

12 The diagram shows part of the graph of $y = 3x\cos 3x$.
Work out the area of the shaded region as an exact value.

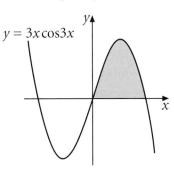

$y = 3x\cos3x$

6 Integration 2

In this chapter you will learn how to find
- the area under a curve defined parametrically
- the volume of a solid of revolution about the x-axis
- the percentage error when the trapezium rule is used to estimate the value of a definite integral

A Area under a curve defined parametrically (answers p 177)

The diagram shows the area under the curve whose equation is $y = 6 - x^2$, between the points where $x = 1$ and $x = 2$.

To find this area you calculate $\int_1^2 (6 - x^2) \, dx$.

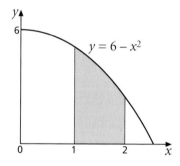

The general form of this integral can be written $\int_a^b y \, dx$, where a and b are the values of x at the end-points.

In the example above, the equation of the curve is given in the explicit form $y = $ a function of x.

In the example on the right, the curve is defined by the parametric equations

$$x = t^2, \qquad y = 1 + \frac{1}{t}$$

The shaded area A is between the points where $t = 0.5$ and $t = 1$.

The integral $\int_a^b y \, dx$, which gives the area, has to be rewritten in terms of t.

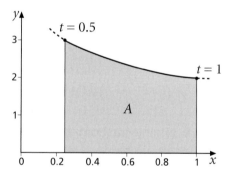

This is like integration by substitution: y is replaced by $1 + \frac{1}{t}$ and dx is replaced by $\frac{dx}{dt} \, dt$, or $2t \, dt$.

The lower and upper limits are the two values of t between which the area lies.

So the area A is $\int_{0.5}^1 \left(1 + \frac{1}{t}\right) 2t \, dt$.

A1 Evaluate the integral above to find the shaded area A.

A2 The diagram shows part of the curve defined by the parametric equations

$$x = t - \frac{1}{t}, \quad y = t^2$$

Find the area under the curve between the points $t = 1$ and $t = 3$.

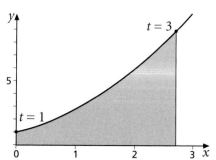

A3 *AB* is part of the curve whose parametric equations are

$$x = t + \frac{1}{t}, \quad y = t - \frac{1}{t}$$

A is the point where $t = 1$. *B* is the point where $t = 2$.

Find the area under the graph between *A* and *B*.

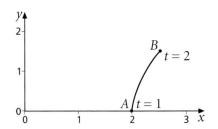

K The area under a curve defined by parametric equations is given by $\int_{t_1}^{t_2} y \, dx$ where y and dx are each expressed in terms of the parameter t, and t_1 and t_2 are the values of t at the end-points of the curve.

Example 1

The diagram shows part of the graph of the function defined by the parametric equations $x = \sin t$, $y = \tan t$. *A* is the point where $t = 0$, *B* the point where $t = \frac{1}{4}\pi$.

Find the area under the graph between *A* and *B*.

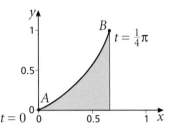

Solution

The area is given by $\int y \, dx$.

$y = \tan t$, $\dfrac{dx}{dt} = \cos t$, so $dx = \cos t \, dt$.

$$\text{Area} = \int_0^{\frac{1}{4}\pi} \tan t \cos t \, dt = \int_0^{\frac{1}{4}\pi} \frac{\sin t}{\cos t} \cos t \, dt = \int_0^{\frac{1}{4}\pi} \sin t \, dt = \left[-\cos t\right]_0^{\frac{1}{4}\pi} = \left(-\frac{1}{\sqrt{2}}\right) - (-1) = 1 - \frac{1}{\sqrt{2}}$$

Exercise A (answers p 177)

1 Find the area under the curve whose parametric equations are $x = t^2$, $y = t + \dfrac{1}{t}$ between the points $t = \frac{1}{2}$ and $t = 2$.

2 In each case below, find the area under the curve whose parametric equations are given, between the points whose values of t are given (in square brackets).

(a) $x = 2(t^2 - 1)$, $y = 3(t + 1)$ $[1, 3]$ (b) $x = 1 + \sqrt{t}$, $y = t(1 + t)$ $[0, 1]$

(c) $x = t + \ln t$, $y = t - \dfrac{1}{t}$ $[1, 2]$ (d) $x = \sin t$, $y = 1 - \sec t$ $[0, \frac{1}{6}\pi]$

(e) $x = \cos 2t$, $y = \sec t$ $[0, \frac{1}{4}\pi]$ (f) $x = t^2$, $y = e^t$ $[0, 2]$

3 The curve shown here is part of the ellipse whose parametric equations are

$$x = 3(1 - \cos t), \quad y = 2 \sin t$$

(a) What is the value of t at the point

  (i) *A*   (ii) *B*

(b) Find the area enclosed by the curve and the *x*-axis.

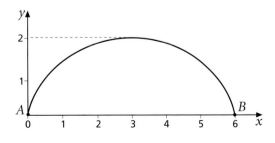

B Volume of a solid revolution about the *x*-axis (answers p 177)

In this diagram the shaded region under a curve is rotated through a complete turn (2π radians) about the *x*-axis to form a **solid of revolution**.

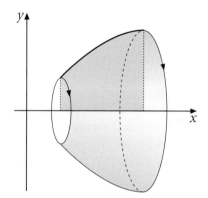

In this section you will learn how to calculate the volume of a solid of revolution when you know the equation of the curve.

The second diagram shows the volume V up to the value x. The radius of the solid at this value is y.

If x is increased by an amount δx, then y is increased by δy and V is increased by δV.

The extra volume δV consists of a slice of thickness δx, whose cross-sectional area varies from a circle of radius y to a circle of radius $(y + \delta y)$.

The volume of the slice lies between $\pi y^2 \delta x$ and $\pi (y + \delta y)^2 \delta x$.

$$\pi y^2 \delta x < \delta V < \pi (y + \delta y)^2 \delta x$$

By dividing by δx, we get

$$\pi y^2 < \frac{\delta V}{\delta x} < \pi (y + \delta y)^2$$

Now suppose that δx gets smaller and smaller.

Then δy gets smaller and smaller and $\dfrac{\delta V}{\delta x}$ gets closer and closer to $\dfrac{dV}{dx}$,

which is 'trapped' between πy^2 and a quantity that gets closer and closer to πy^2.

So $\dfrac{dV}{dx} = \pi y^2$ and this is equivalent to $V = \int \pi y^2 \, dx$.

To find the volume between two given values of x, you calculate a definite integral:

The volume of a solid of revolution about the *x*-axis between $x = a$ and $x = b$

is given by $\displaystyle\int_{a}^{b} \pi y^2 \, dx$.

B1 A solid of revolution is formed by rotating about the x-axis the region under the line $y = \frac{1}{2}x$ between $x = 0$ and $x = 3$.

(a) What name is given to this solid of revolution?

(b) Express y^2 in terms of x.

(c) By substituting for y^2 in $\int_0^3 \pi y^2 \, dx$, find the volume of the solid of revolution.

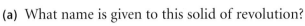

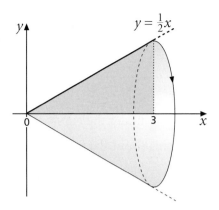

B2 The sloping line in this diagram goes through the point (h, r). When the shaded area is rotated through 2π radians about the x-axis, the solid of revolution formed is a cone of base radius r and height h.

(a) Explain why the equation of the line is $y = \dfrac{r}{h}x$.

(b) By substituting for y^2 in $\int_0^h \pi y^2 \, dx$, show that the volume of the cone is $\frac{1}{3}\pi r^2 h$.

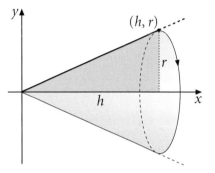

B3 A solid of revolution is formed by rotating about the x-axis the region under the curve $y = x^2$ between $x = 1$ and $x = 2$.

(a) Express y^2 in terms of x.

(b) Find the volume of the solid of revolution.

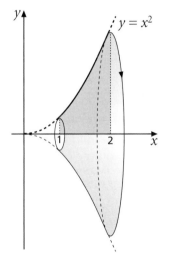

B4 This diagram shows the circle of radius r whose centre is at the origin. The equation of the circle is $x^2 + y^2 = r^2$, which can be written as $y^2 = r^2 - x^2$.

(a) What solid is formed when the shaded region is rotated through 2π radians about the x-axis?

(b) The volume of this solid is given by $\int_{-r}^{r} \pi y^2 \, dx$.

Using the fact that $y^2 = r^2 - x^2$, show that the volume of the solid of revolution is $\frac{4}{3}\pi r^3$.

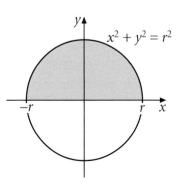

The integration that you need to do in order to find the volume of a solid of revolution may require one of the techniques you met in the previous chapter.

Example 2

The region under the curve $y = \sqrt{\dfrac{1+x}{x}}$ between $x = 1$ and $x = 4$ is rotated through 2π radians about the x-axis to form a solid of revolution. Find the volume of the solid.

Solution

$$\int_1^4 \pi y^2 \, dx = \int_1^4 \pi \left(\frac{1+x}{x}\right) dx = \pi \int_1^4 \left(\frac{1}{x} + \frac{x}{x}\right) dx = \pi \int_1^4 \left(\frac{1}{x} + 1\right) dx$$

$$= \pi \left[\ln|x| + x\right]_1^4 = \pi((\ln 4 + 4) - (0 + 1)) = \pi(\ln 4 + 3)$$

Example 3

The region under the curve $y = \sqrt{\dfrac{2x}{x^2 + 3}}$ between $x = 0$ and $x = 1$ is rotated through 2π radians about the x-axis to form a solid of revolution. Find the volume of the solid.

Solution

$$\text{Volume} = \int_0^1 \pi y^2 \, dx = \pi \int_0^1 \frac{2x}{x^2 + 3} \, dx \qquad \left(\textit{This integral is of the form } \int \frac{f'(x)}{f(x)} \, dx.\right)$$

$$= \pi \left[\ln|x^2 + 3|\right]_0^1 = \pi(\ln 4 - \ln 3) = \pi \ln \tfrac{4}{3}$$

Exercise B (answers p 177)

1 The region under the curve $y = \sqrt{x(4-x)}$ between $x = 0$ and $x = 4$ is rotated through 2π radians about the x-axis.
Find the volume of the solid of revolution formed.

2 The region under the curve $y = 1 - x^2$ between $x = 0$ and $x = 1$ is rotated through 2π radians about the x-axis.
Find the volume of the solid of revolution formed.

3 Find the volume of the solid of revolution formed when the region enclosed by each of these curves and the x-axis, between the given values of x, is rotated through 2π radians about the x-axis.

(a) $y = e^x$ from $x = 0$ to $x = 3$

(b) $y = \dfrac{1}{x^2}$ from $x = 1$ to $x = 4$

(c) $y = x + \dfrac{1}{x}$ from $x = 1$ to $x = 3$

(d) $y = \sqrt{\sin 3x}$ from $x = 0$ to $x = \tfrac{1}{3}\pi$

(e) $y = \sqrt{x \sin 2x}$ from $x = 0$ to $x = \tfrac{1}{2}\pi$

(f) $y = \sqrt{x e^{-x}}$ from $x = 0$ to $x = 2$

(g) $y = \dfrac{1}{\sqrt{1 + 2x}}$ from $x = 0$ to $x = 1$

(h) $y = \sqrt{\dfrac{x}{x^2 + 1}}$ from $x = 0$ to $x = 3$

C Solid of revolution defined parametrically (answers p 178)

If the curve that is rotated is defined parametrically, the volume is still $\int \pi y^2 \, dx$, but you need to substitute for y, for dx and for the limits of integration.

For example, here is the part of the curve whose parametric equations are

$$x = t^2, \qquad y = 1 + 2t$$

The shaded area A between the points where $t = 0$ and $t = 1$ is rotated through 2π radians about the x-axis.

The volume is given by $\int_0^1 \pi y^2 \, dx$.

$$y^2 = (1 + 2t)^2, \quad dx = 2t \, dt, \quad \text{so} \quad \int_0^1 \pi y^2 \, dx = \pi \int_0^1 (1 + 2t)^2 \, 2t \, dt$$

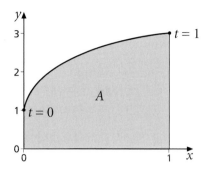

C1 Show that the value of the integral above is $\dfrac{17\pi}{3}$.

Example 4

The region between $t = 0$ and $t = \frac{1}{4}\pi$ under the curve with parametric equations $x = \tan t$, $y = \sin t$ is rotated through 2π radians about the x-axis.

Find the volume of the solid of revolution formed.

Solution

$y^2 = \sin^2 t, \quad dx = \sec^2 t \, dt$

$$\text{Volume} = \int_0^{\frac{1}{4}\pi} \pi y^2 \, dx = \pi \int_0^{\frac{1}{4}\pi} \sin^2 t \sec^2 t \, dt$$

$$= \pi \int_0^{\frac{1}{4}\pi} \frac{\sin^2 t}{\cos^2 t} \, dt = \pi \int_0^{\frac{1}{4}\pi} \tan^2 t \, dt = \pi \int_0^{\frac{1}{4}\pi} \left(\sec^2 t - 1 \right) dt$$

$$= \pi \left[\tan t - t \right]_0^{\frac{1}{4}\pi} = \pi \left(1 - \tfrac{1}{4}\pi \right)$$

Exercise C (answers p 178)

1 Find the volume of the solid of revolution formed when the region between $t = 0$ and $t = 2$ under the curve with parametric equations $x = 2t^2$, $y = 3(t - 1)$ is rotated through 2π radians about the x-axis.

2 In each case below, find the volume of the solid of revolution formed when the region under the curve with the given parametric equations, between the given values of the parameter (in square brackets), is rotated through 2π radians about the x-axis.

(a) $x = t^2 + 1$, $y = t^3 - 1$ $[0, 1]$ (b) $x = t + \ln t$, $y = t^2$ $[1, e]$

(c) $x = \sqrt{t} + t$, $y = \sqrt{t} - t$ $[0, 1]$ (d) $x = \sin \theta$, $y = \sqrt{\sin \theta + \cos \theta}$ $[0, \frac{1}{2}\pi]$

3 Find the volume of the solid of revolution formed when the region between the points $t = 1$ and $t = 2$ under the curve with parametric equations $x = t + \dfrac{1}{\sqrt{t}}$, $y = t - \dfrac{1}{\sqrt{t}}$ is rotated through 2π radians about the x-axis.

4 The region between the points $t = 0$ and $t = 2$ under the curve with parametric equations $x = t^2$, $y = e^t$ is rotated through 2π radians about the x-axis. Find the volume of the solid of revolution formed.

5 Find the volume of the solid of revolution formed when the region between $\theta = \frac{1}{4}\pi$ and $\theta = \frac{1}{3}\pi$ under the curve with parametric equations $x = \ln \sin \theta$, $y = \tan \theta$ is rotated through 2π radians about the x-axis.

6 (a) Show that the derivative of $\ln(\sec x + \tan x)$ is $\sec x$.

(b) The region between $\theta = 0$ and $\theta = \frac{1}{4}\pi$ under the curve with parametric equations $x = \sin \theta$, $y = \tan \theta$ is rotated through 2π radians about the x-axis. Show that the volume of the solid of revolution formed is given by

$$\pi \int_0^{\frac{1}{4}\pi} (\sec \theta - \cos \theta)\, d\theta$$

(c) Show that the value of this integral is $\pi\left[\ln\left(1 + \sqrt{2}\right) - \tfrac{1}{2}\sqrt{2}\right]$.

D Trapezium rule: percentage error (answers p 178)

The function $\sqrt{1 + e^x}$ cannot be integrated by the methods you have met so far.

However, the trapezium rule can be used to estimate a definite integral such as

$\int_0^2 \sqrt{1 + e^x}\, dx$.

The trapezium rule is obtained by replacing the graph of the function by a set of straight line segments, as shown in the diagram.

The total area of the trapezia formed in this way is an estimate of the area under the graph.

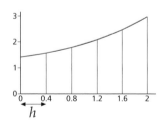

Estimate $= \frac{1}{2}h\left[\text{sum of end ordinates} + 2(\text{sum of other ordinates})\right]$

D1 (a) Copy and complete this table of values for the function $y = \sqrt{1 + e^x}$.

x	0	0.4	0.8	1.2	1.6	2
y	1.414	1.579				

(b) Use the trapezium rule with six ordinates to find an estimate of $\int_0^2 \sqrt{1 + e^x}\, dx$.

(c) How can you tell from the graph that the result in (b) is an overestimate of the value of the integral?

(d) How could you improve on the estimate?

If the trapezium rule is used for a function that can be integrated, the exact value of the integral can be found and hence the percentage error in the estimate.

Example 5

The region under the curve $y = e^{-x}$ between $x = 0$ and $x = 1$ is rotated through 2π radians about the x-axis to form a solid of revolution.

(a) Use the trapezium rule with three ordinates to estimate the volume of the solid.

(b) Find by integration the exact value of the volume.

(c) Find the percentage error in the estimate in (a).

Solution

The volume is given by $\int_0^1 \pi y^2 \, dx = \pi \int_0^1 e^{-2x} \, dx$

(a) Three ordinates means two strips, so $h = 0.5$. Values of e^{-2x} are:

x	0	0.5	1
e^{-2x}	1	0.3679	0.1353

Estimate $= \pi \times 0.25 \times (1 + 0.1353 + 2 \times 0.3679) = 1.47$ (to 3 s.f.)

(b) Volume $= \pi \int_0^1 e^{-2x} \, dx = \pi \left[-\frac{1}{2} e^{-2x} \right]_0^1 = -\frac{1}{2}\pi \left(e^{-2} - 1 \right) = \frac{1}{2}\pi \left(1 - e^{-2} \right)$

(c) Numerical value of volume $= \frac{1}{2}\pi(1 - e^{-2}) = 1.36$ (to 3 s.f.)

Actual error $= 1.469... - 1.358... = 0.111...$ Percentage error $= \frac{0.111...}{1.358...} \times 100 = 8.2\%$ (to 1 d.p.).

Exercise D (answers p 178)

1 Use the trapezium rule with six ordinates to estimate the value of $\int_1^5 \sqrt{\ln x} \, dx$.

2 (a) Use the trapezium rule with three ordinates to estimate the value of $\int_0^2 \sqrt{4 - x^2} \, dx$.

(b) Obtain a better estimate by using five ordinates.

(c) Use the substitution $x = 2\sin\theta$ to find the exact value of the integral.

(d) Find the percentage error in the estimate obtained by using the trapezium rule with
  (i) 3 ordinates   (ii) 5 ordinates
  Comment on the results.

3 (a) Use the trapezium rule with three ordinates to estimate the value of $\int_1^3 x \ln x \, dx$.

(b) Obtain a better estimate by using five ordinates.

(c) Use integration by parts to show that the exact value of the integral is $\frac{9}{2} \ln 3 - 2$.

(d) Find the percentage error in the estimate obtained by using the trapezium rule with
  (i) 3 ordinates   (ii) 5 ordinates
  Comment on the results.

4 The region under the graph of $y = \ln x$ between $x = 1$ and $x = 3$ is rotated through 2π radians about the x-axis to form a solid of revolution.

(a) Use the trapezium rule with five ordinates to estimate the volume of the solid.

(b) By using the substitution $x = e^u$, find the exact value of the volume.

(c) Find the percentage error in the estimate of the volume.

Key points

- The area under a curve defined by parametric equations is given by $\int_{t_1}^{t_2} y \, dx$
  where y and dx are each expressed in terms of the parameter t, and
  t_1 and t_2 are the values of t at the end-points of the curve. (p 79)

- The volume of a solid of revolution about the x-axis between $x = a$ and $x = b$
  is given by $\int_{a}^{b} \pi y^2 \, dx$.

  If x and y are defined in terms of a parameter t, then the volume is $\int_{t_1}^{t_2} \pi y^2 \, dx$,
  where y and dx are each expressed in terms of the parameter t. (pp 80, 83)

- The trapezium rule gives an estimate of the value of a definite integral:
  $\frac{1}{2}h\big[$sum of end ordinates + 2(sum of other ordinates)$\big]$, where h is the
  gap between ordinates. (p 84)

Mixed questions (answers p 179)

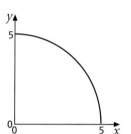

1 The parametric equations for the quarter-circle of radius 5
between $(0, 5)$ and $(5, 0)$ may be written as
$$x = 5 \sin \theta, \quad y = 5 \cos \theta \quad \left(0 \le \theta \le \tfrac{1}{2}\pi\right)$$
Show by integration that the area of the quarter-circle is $\dfrac{25\pi}{4}$.

2 The region R bounded by the curve $y = x^{\frac{1}{2}}e^{3x}$, the x-axis and the lines
$x = 1$ and $x = 4$ is rotated through 2π radians about the x-axis.
Find the volume of the solid of revolution formed.

3 The region between $x = 0$ and $x = 3$ under the curve $y = \dfrac{1}{\sqrt{9 + x^2}}$ is
rotated through 2π radians about the x-axis to form a solid of revolution.

  (a) Use the trapezium rule with four ordinates to estimate the volume of the solid.

  (b) By using the substitution $x = 3 \tan \theta$, show that the exact value of the volume
  is $\frac{1}{12}\pi^2$.

  (c) Find the percentage error in the estimated volume.

4 A curve has the parametric equations $x = \theta + \tan \theta$, $y = \cos \theta$.

  The region under the curve between $\theta = 0$ and $\theta = \frac{1}{4}\pi$ is rotated through
  2π radians about the x-axis to form a solid of revolution.

  Show that the volume of the solid is $\dfrac{\pi}{4} + \dfrac{3\pi^2}{8}$.

5 (a) Use integration by parts to show that $\int_{0}^{1} x^2 e^{-x} \, dx = 2 - 5e^{-1}$.

  (b) Find the volume of the solid of revolution formed when the region
  under the curve $y = (1 + x)e^{-\frac{1}{2}x}$ between $x = 0$ and $x = 1$ is rotated
  through 2π radians about the x-axis.

6 *ABC* is part of the curve whose parametric equations are $x = \tan\theta$, $y = 1 - \sin\theta$.

A is the point where $\theta = 0$, *B* is the point where $\theta = \frac{1}{6}\pi$ and *C* is the point where $\theta = \frac{1}{4}\pi$.

(a) Estimate the area under the curve by replacing the curve *ABC* by two line segments *AB* and *BC*.

(b) Find the exact value of the area.

(c) Find the percentage error in the estimate.

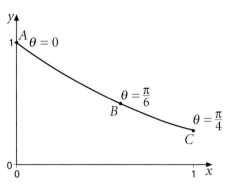

Test yourself (answers p 179)

1 When a wheel of unit radius rolls along horizontal ground, the locus of a point on its circumference is a curve called a 'cycloid'.

The parametric equations of this curve can be written

$$x = \theta - \sin\theta, \quad y = 1 - \cos\theta$$

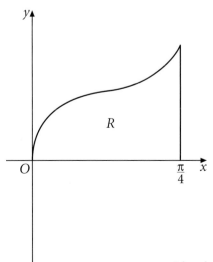

Find the area under the curve between the points where $\theta = 0$ and $\theta = 2\pi$.

2 The region between the points $t = 0$ and $t = 1$ under the curve whose parametric equations are $x = \ln t$, $y = te^t$ is rotated through 2π radians about the *x*-axis. Find the volume of the solid of revolution generated.

3 (a) Use the trapezium rule with five ordinates to estimate the area under the curve $y = \dfrac{1}{1 + \sqrt{x}}$ between $x = 0$ and $x = 1$.

(b) Find the exact value of the area by integration, using the substitution $u = 1 + \sqrt{x}$ or otherwise.

(c) Find the percentage error in the estimate in part (a).

4 (a) Use integration by parts to show that

$$\int_0^{\frac{\pi}{4}} x \sec^2 x \, dx = \tfrac{1}{4}\pi - \tfrac{1}{2}\ln 2$$

The finite region *R*, bounded by the curve with equation $y = x^{\frac{1}{2}}\sec x$, the line $x = \dfrac{\pi}{4}$ and the *x*-axis, is shown in the diagram.

The region *R* is rotated through 2π radians about the *x*-axis.

(b) Find the volume of the solid of revolution generated.

(c) Find the gradient of the curve with equation $y = x^{\frac{1}{2}}\sec x$ at the point where $x = \dfrac{\pi}{4}$.

Edexcel

7 Integration 3

In this chapter you will learn how to integrate a rational function using partial fractions

A Using partial fractions (answers p 180)

The process of expressing a rational expression in a form involving partial fractions was introduced in chapter 1. There are two cases to consider:

    (1) degree of numerator < degree of denominator (a 'proper' algebraic fraction)

    (2) degree of numerator ≥ degree of denominator (an 'improper' algebraic fraction)

Example 1 below illustrates case (1).

Example 1

Express $\dfrac{6x^2 + x + 1}{(x-1)(x+1)^2}$ as partial fractions.

Solution

The degree of the numerator is 2. This is less than that of the denominator, which is 3.

There will be three partial fractions, with denominators $(x-1)$, $(x+1)$ and $(x+1)^2$.

Let $\dfrac{6x^2 + x + 1}{(x-1)(x+1)^2} \equiv \dfrac{A}{x-1} + \dfrac{B}{x+1} + \dfrac{C}{(x+1)^2}$.

So $\dfrac{6x^2 + x + 1}{(x-1)(x+1)^2} \equiv \dfrac{A(x+1)^2 + B(x+1)(x-1) + C(x-1)}{(x-1)(x+1)^2}$.

So $A(x+1)^2 + B(x-1)(x+1) + C(x-1) \equiv 6x^2 + x + 1$.

In this identity, let $x = -1$ (so that A and B are eliminated): $-2C = 6 - 1 + 1$, so $C = -3$.

Now let $x = 1$ (so that B and C are eliminated): $4A = 6 + 1 + 1$, so $A = 2$.

With A and C now known, you can let x be any other value in order to find B.

Alternatively, you can look at the coefficient of x^2 on each side of the identity.

On the left the coefficient of x^2 is $A + B$; on the right it is 6. So $A + B = 6$, from which $B = 4$.

So $\dfrac{6x^2 + x + 1}{(x-1)(x+1)^2} \equiv \dfrac{2}{x-1} + \dfrac{4}{x+1} - \dfrac{3}{(x+1)^2}$.

A1 Express each of these as partial fractions.

    (a) $\dfrac{5x-4}{(x-2)(x+1)}$ (b) $\dfrac{2x+11}{(x-2)(2x+1)}$ (c) $\dfrac{5x+13}{(x-3)(x+1)^2}$

In case (2), where the degree of the numerator is greater than or equal to that of the denominator, there will be a **quotient** as well as partial fractions. The degree of the quotient is equal to the difference between the degrees of the numerator and denominator.

degree 2

$$\frac{2x^2 + x + 5}{x - 3} \equiv Ax + B + \frac{C}{x - 3}$$

degree 1 $2 - 1 = 1$,
                so quotient is **linear**

degree 2

$$\frac{x^2 + 4}{(x - 2)(x + 5)} \equiv A + \frac{B}{x - 2} + \frac{C}{x + 5}$$

degree 2 $2 - 2 = 0$,
                so quotient is a **constant**

Example 2

Express $\dfrac{x^2 + 5}{(x - 1)(x + 2)}$ in a form involving partial fractions.

Solution

The degree of the numerator is 2; that of the denominator is also 2. So there will be a quotient (a constant).

There are two possible methods.

Method 1

Let $\dfrac{x^2 + 5}{(x - 1)(x + 2)} \equiv A + \dfrac{B}{x - 1} + \dfrac{C}{x + 2}$.

So $A(x - 1)(x + 2) + B(x + 2) + C(x - 1) \equiv x^2 + 5$.

Let $x = 1$. Then $3B = 6$, so $B = 2$.

Let $x = -2$. Then $-3C = 9$, so $C = -3$.

To find the value of A, choose another value for x, or equate coefficients of x^2, giving $A = 1$.

So $\dfrac{x^2 + 5}{(x - 1)(x + 2)} \equiv 1 + \dfrac{2}{x - 1} - \dfrac{3}{x + 2}$.

Method 2

$$\frac{x^2 + 5}{(x - 1)(x + 2)} \equiv \frac{x^2 + 5}{x^2 + x - 2}$$

$$\equiv \frac{(x^2 + x - 2) + 7 - x}{x^2 + x - 2}$$

$$\equiv 1 + \frac{7 - x}{x^2 + x - 2}$$

$$\equiv 1 + \frac{7 - x}{(x - 1)(x + 2)}$$

Let $\dfrac{7 - x}{(x - 1)(x + 2)} \equiv \dfrac{A}{x - 1} + \dfrac{B}{x + 2}$.

Then $A(x + 2) + B(x - 1) \equiv 7 - x$

Let $x = -2$. Then $-3B = 9$, so $B = -3$.

Let $x = 1$. Then $3A = 6$, so $A = 2$.

So $\dfrac{x^2 + 5}{(x - 1)(x + 2)} \equiv 1 + \dfrac{2}{x - 1} - \dfrac{3}{x + 2}$.

A2 Express each of these in a form involving partial fractions.

(a) $\dfrac{x^2 + 3}{(x + 1)(x + 2)}$

(b) $\dfrac{x^2}{(x - 2)(x + 3)}$

(c) $\dfrac{2x^2 - 3}{(2 - x)(x + 1)}$

Partial fractions can be used to help integrate a function.

For example, the function $\dfrac{x+9}{(x-3)(x+1)}$ is equivalent to $\dfrac{3}{x-3} - \dfrac{2}{x+1}$.

It follows that $\displaystyle\int \dfrac{x+9}{(x-3)(x+1)}\,dx = 3\int \dfrac{1}{x-3}\,dx - 2\int \dfrac{1}{x+1}\,dx$.

The integrals on the right side above are of a type you have met before: $\displaystyle\int \dfrac{1}{x+b}\,dx = \ln|x+b| + c$.

So $\displaystyle\int \dfrac{x+9}{(x-3)(x+1)}\,dx = 3\ln|x-3| - 2\ln|x+1| + c$.

A3 Use your answer to A1 (a) to find $\displaystyle\int \dfrac{5x-4}{(x-2)(x+1)}\,dx$.

A more general type of integral that you have met before is $\displaystyle\int \dfrac{1}{ax+b}\,dx = \dfrac{1}{a}\ln|ax+b| + c$.

A4 Use your answer to A1 (b) to find $\displaystyle\int \dfrac{2x+11}{(x-2)(2x+1)}\,dx$.

A5 (a) Express $\dfrac{x-6}{x(x+2)}$ as partial fractions.

(b) Find $\displaystyle\int \dfrac{x-6}{x(x+2)}\,dx$.

A6 Use your answers to question A2 to find

(a) $\displaystyle\int \dfrac{x^2+3}{(x+1)(x+2)}\,dx$ **(b)** $\displaystyle\int \dfrac{x^2}{(x-2)(x+3)}\,dx$ **(c)** $\displaystyle\int \dfrac{2x^2-3}{(2-x)(x+1)}\,dx$

In example 1 it was shown that $\dfrac{6x^2+x+1}{(x-1)(x+1)^2} \equiv \dfrac{2}{x-1} + \dfrac{4}{x+1} - \dfrac{3}{(x+1)^2}$.

It follows that $\displaystyle\int \dfrac{6x^2+x+1}{(x-1)(x+1)^2}\,dx = 2\int \dfrac{1}{x-1}\,dx + 4\int \dfrac{1}{x+1}\,dx - 3\int \dfrac{1}{(x+1)^2}\,dx$.

The first two integrals on the right are of the same type as before, but the third is different.

$\displaystyle\int \dfrac{1}{(x+1)^2}\,dx$ is the same as $\displaystyle\int (x+1)^{-2}\,dx$, which is $\dfrac{(x+1)^{-1}}{-1} + c = -\dfrac{1}{x+1} + c$.

A7 Complete the working above to find $\displaystyle\int \dfrac{6x^2+x+1}{(x-1)(x+1)^2}\,dx$.

A8 Use your answer to A1 (c) to find $\displaystyle\int \dfrac{5x+13}{(x-3)(x+1)^2}\,dx$.

Example 3

(a) Express $\dfrac{x+5}{(2x+1)(x-1)^2}$ as partial fractions. **(b)** Hence find $\displaystyle\int\dfrac{x+5}{(2x+1)(x-1)^2}\,dx$.

Solution

(a) Let $\dfrac{x+5}{(2x+1)(x-1)^2} \equiv \dfrac{A}{2x+1} + \dfrac{B}{x-1} + \dfrac{C}{(x-1)^2} \equiv \dfrac{A(x-1)^2 + B(2x+1)(x-1) + C(2x+1)}{(2x+1)(x-1)^2}$.

So $A(x-1)^2 + B(2x+1)(x-1) + C(2x+1) \equiv x+5$.

Let $x = 1$. Then $3C = 6$, so $C = 2$.

Let $x = -\tfrac{1}{2}$. Then $\tfrac{9}{4}A = 4\tfrac{1}{2} = \tfrac{9}{2}$, so $A = 2$.

Equate coefficients of x^2: $A + 2B = 0$, so $B = -\tfrac{1}{2}A = -1$.

So $\dfrac{x+5}{(2x+1)(x-1)^2} \equiv \dfrac{2}{2x+1} - \dfrac{1}{x-1} + \dfrac{2}{(x-1)^2}$.

(b) $\displaystyle\int\dfrac{x+5}{(2x+1)(x-1)^2}\,dx = 2\int\dfrac{1}{2x+1}\,dx - \int\dfrac{1}{x-1}\,dx + 2\int\dfrac{1}{(x-1)^2}\,dx$

$$= 2 \times \tfrac{1}{2}\ln|2x+1| - \ln|x-1| - \dfrac{2}{x-1} + c$$

$$= \ln|2x+1| - \ln|x-1| - \dfrac{2}{x-1} + c$$

Example 4

Find $\displaystyle\int\dfrac{x^2}{(x-2)(x+1)}\,dx$.

Solution

The function here is an improper fraction. The degrees of numerator and denominator are both 2, so there is a constant quotient + partial fractions.

Let $\dfrac{x^2}{(x-2)(x+1)} \equiv A + \dfrac{B}{x-2} + \dfrac{C}{x+1}$.

So $A(x-2)(x+1) + B(x+1) + C(x-2) \equiv x^2$.

Let $x = 2$. Then $3B = 4$, so $B = \tfrac{4}{3}$.

Let $x = -1$. Then $-3C = 1$, so $C = -\tfrac{1}{3}$.

By equating coefficients of x^2, $A = 1$.

So $\displaystyle\int\dfrac{x^2}{(x-2)(x+1)}\,dx = \int 1\,dx + \tfrac{4}{3}\int\dfrac{1}{x-2}\,dx - \tfrac{1}{3}\int\dfrac{1}{x+1}\,dx$

$$= x + \tfrac{4}{3}\ln|x-2| - \tfrac{1}{3}\ln|x+1| + c$$

Exercise A (answers p 180)

1 Find each of these indefinite integrals.

(a) $\displaystyle\int\frac{1}{x(x-1)}\,dx$

(b) $\displaystyle\int\frac{3}{(x-1)(x+2)}\,dx$

(c) $\displaystyle\int\frac{x}{(x-1)(x+3)}\,dx$

(d) $\displaystyle\int\frac{x+5}{(x-1)(x+4)}\,dx$

(e) $\displaystyle\int\frac{x+7}{(2x-1)(x+2)}\,dx$

(f) $\displaystyle\int\frac{5x+1}{(1-x)(2x+1)}\,dx$

2 Find each of these indefinite integrals.

(a) $\displaystyle\int\frac{x^2}{(x+1)(x+2)}\,dx$

(b) $\displaystyle\int\frac{x^2+3}{x(x-3)}\,dx$

(c) $\displaystyle\int\frac{x^2-3}{(x+2)(x-1)}\,dx$

3 Find each of these indefinite integrals.

(a) $\displaystyle\int\frac{x}{(x-2)(x+1)^2}\,dx$

(b) $\displaystyle\int\frac{x+3}{(x+1)(x-1)^2}\,dx$

(c) $\displaystyle\int\frac{x}{(2x-1)(x-1)^2}\,dx$

(d) $\displaystyle\int\frac{x-1}{x^2(x+1)}\,dx$

(e) $\displaystyle\int\frac{x+1}{(x+2)^2(2x+3)}\,dx$

(f) $\displaystyle\int\frac{x-1}{x(x-4)^2}\,dx$

B Definite integrals

As with all integrals involving expressions such as $\dfrac{1}{x}, \dfrac{1}{x-2}, \dfrac{1}{2x+3}, \dfrac{1}{(x-2)^2}, \ldots$

you need to make sure that the range of a definite integral does not include any discontinuity.

For example, the definite integral $\displaystyle\int_1^4\frac{1}{x-2}\,dx$

cannot be evaluated because between $x=1$ and $x=4$ there is a discontinuity at $x=2$.

In all the examples and questions that follow, this problem will not arise.

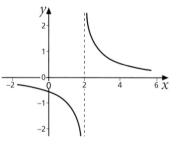

Example 5

Find $\displaystyle\int_2^3\frac{x+5}{(2x+1)(x-1)^2}\,dx$.

Give your answer in the form $a+\ln b$.

Solution

The indefinite integral has been found already, in example 3.

$$\int_2^3\frac{x+5}{(2x+1)(x-1)^2}\,dx=\left[\ln|2x+1|-\ln|x-1|-\frac{2}{x-1}\right]_2^3$$

$$=(\ln 7-\ln 2-1)-(\ln 5-\ln 1-2)$$

$$=1+\ln 7-\ln 2-\ln 5$$

$$=1+\ln\left(\frac{7}{2\times 5}\right)\quad \textit{from the laws of logarithms}$$

$$=1+\ln\tfrac{7}{10}$$

Exercise B (answers p 180)

1 Find the exact value of each of these.

(a) $\int_1^2 \dfrac{1}{x(x+1)} \, dx$

(b) $\int_0^3 \dfrac{1}{(x+1)(x+4)} \, dx$

(c) $\int_1^4 \dfrac{x+1}{x(x-5)} \, dx$

(d) $\int_0^1 \dfrac{x+2}{(x+1)(x+5)} \, dx$

(e) $\int_3^4 \dfrac{x+3}{(x-2)(2x-1)} \, dx$

(f) $\int_0^1 \dfrac{x+6}{(x+1)(2x-3)} \, dx$

2 Find the exact value of each of these.

(a) $\int_1^2 \dfrac{x^2+3}{x(x+1)} \, dx$

(b) $\int_2^5 \dfrac{x^2}{(x-1)(x+2)} \, dx$

(c) $\int_0^2 \dfrac{x^2-1}{(x+2)(x+4)} \, dx$

(d) $\int_0^2 \dfrac{x^2}{(x+5)(x-3)} \, dx$

(e) $\int_{-1}^1 \dfrac{x^2+2}{(2-x)(x+4)} \, dx$

(f) $\int_0^1 \dfrac{x^2}{x^2-4} \, dx$

3 Find the exact value of each of these.

(a) $\int_0^2 \dfrac{x}{(x+1)(x-3)^2} \, dx$

(b) $\int_2^5 \dfrac{x+1}{(x-1)(x+2)^2} \, dx$

(c) $\int_1^3 \dfrac{x+4}{x^2(x+2)} \, dx$

(d) $\int_0^1 \dfrac{x}{(x+1)^2(x+2)} \, dx$

(e) $\int_0^2 \dfrac{x+2}{(2x+1)(x+1)^2} \, dx$

(f) $\int_0^2 \dfrac{x+2}{(x+3)^2(2x+5)} \, dx$

4 Show that $\displaystyle\int_0^{\frac{1}{3}} \dfrac{4}{(1+3x)(1-x)} \, dx = \ln 3$.

5 Show that $\displaystyle\int_{-\frac{1}{2}}^{\frac{1}{2}} \dfrac{9}{(1+x)(2-x)^2} \, dx = \frac{4}{5} + \ln 5$.

6 Evaluate $\displaystyle\int_1^2 \dfrac{2x^2-11x+19}{(1+x)(3-x)^2} \, dx$, giving your answer in the form $a + \ln b$.

Key points

- Functions such as $\dfrac{x^2}{(x-2)(x+3)}$, $\dfrac{x+5}{(2x+1)(x-1)^2}$ and so on can be integrated by using partial fractions.
  If degree of numerator $\geq$ degree of denominator, then there will be a quotient + partial fractions. (pp 88–91)

Mixed questions (answers p 181)

1 (a) Express $\dfrac{3(x+1)}{x(x+3)}$ as partial fractions.

(b) Hence find $\displaystyle\int_1^4 \dfrac{3(x+1)}{x(x+3)} \, dx$.

2 The function f is given by $f(x) = \dfrac{x+2}{(1-x)(x+5)}$, $-5 < x < 1$.

(a) Given that $f(x) = \dfrac{A}{1-x} + \dfrac{B}{x+5}$ find the values of the constants A and B.

(b) Evaluate $\displaystyle\int_{-2}^{0} f(x)\,dx$, giving the exact answer in the form $\ln(a)$, where a is a constant to be found.

3 (a) Given that $\dfrac{2x^2 - 5}{(x+1)(2x-1)} = A + \dfrac{B}{x+1} + \dfrac{C}{2x-1}$ find the values of the constants A, B and C.

(b) Hence find $\displaystyle\int \dfrac{2x^2 - 5}{(x+1)(2x-1)}\,dx$.

(c) Show that $\displaystyle\int_{1}^{2} \dfrac{2x^2 - 5}{(x+1)(2x-1)}\,dx = 1 - \ln(2\sqrt{3})$.

4 $g(x) = \dfrac{9x - 1}{(1+x)^2(2x-3)}$, $x > \frac{3}{2}$.

(a) Express $g(x)$ in the form $\dfrac{A}{(x+1)^2} + \dfrac{B}{x+1} + \dfrac{C}{2x-3}$, where A, B and C are constants to be found.

(b) Find $\displaystyle\int_{2}^{3} g(x)\,dx$ in the form $p + q\ln\frac{3}{2}$, where p and q are rational constants.

5 The diagram shows part of the curve with equation $y = f(x)$, where

$$f(x) = \dfrac{14x + 9}{(x+1)^2(3x+2)}, \quad x \geq 0$$

The finite region R is bounded by the curve, the coordinate axes and the line $x = 4$.

Find the area of R, giving your answer in the form $a + \ln b$ where a and b are rational constants to be found.

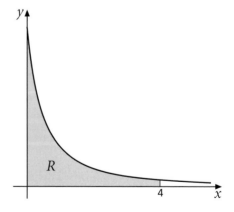

6 The finite region R is bounded by the curve with equation $y = \dfrac{13x + 11}{(1+8x)(3-x)}$, the coordinate axes and the line $x = \frac{1}{2}$.

Find the area of R, giving your answer in the form $a\ln 5 + b\ln 6$.

*7 Find each of these indefinite integrals.
You may need to use methods from earlier chapters on integration.

(a) $\displaystyle\int \frac{6}{x^2-1}\,dx$ (b) $\displaystyle\int \frac{2x}{x^2-1}\,dx$ (c) $\displaystyle\int \frac{x^2+1}{x(x-5)}\,dx$ (d) $\displaystyle\int \frac{2x-5}{x(x-5)}\,dx$

(e) $\displaystyle\int \frac{x}{x^2+1}\,dx$ (f) $\displaystyle\int \frac{6}{x^2+1}\,dx$ (g) $\displaystyle\int \frac{x^2}{x^2+1}\,dx$ (h) $\displaystyle\int \frac{\ln x}{(x-1)^2}\,dx$

Test yourself (answers p 181)

1 Find

(a) $\displaystyle\int \frac{x+1}{x(3x+2)}\,dx$ (b) $\displaystyle\int \frac{x^2+1}{x(3x+2)}\,dx$ (c) $\displaystyle\int \frac{x+1}{x^2(3x+2)}\,dx$

2 Evaluate each of these.

(a) $\displaystyle\int_1^2 \frac{x}{(3x-1)(x+1)}\,dx$ (b) $\displaystyle\int_1^2 \frac{x^2}{(3x-1)(x+1)}\,dx$ (c) $\displaystyle\int_1^2 \frac{x-1}{(3x-1)(x+1)^2}\,dx$

3 The diagram shows part of the curve with equation $y = f(x)$, where

$$f(x) = \frac{x^2+4}{(1+x)(4-x)}, \quad 0 \le x < 4$$

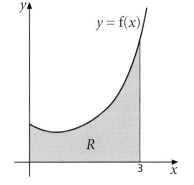

(a) Given that $f(x) = A + \dfrac{B}{1+x} + \dfrac{C}{4-x}$, find the values of the constants A, B and C.

(b) The finite region R is bounded by the curve, the coordinates axes and the line $x = 3$.
Find the area of R, giving your answer in the form $p + q\ln 2$, where p and q are integers to be found.

4 Given that $\dfrac{26x+1}{(3-2x)^2(x+1)} \equiv \dfrac{A}{(3-2x)^2} + \dfrac{B}{3-2x} + \dfrac{C}{x+1}$,

(a) find the values of A, B and C

(b) find the exact value of $\displaystyle\int_0^{\frac{2}{3}} \frac{26x+1}{(3-2x)^2(x+1)}\,dx$

5 Show that $\displaystyle\int_0^{\frac{1}{3}} \frac{2x+1}{(1-2x)^2(x+1)}\,dx = \frac{4}{3} - \frac{2}{9}\ln 2$.

8 Differential equations

In this chapter you will learn how to
- form a first-order differential equation
- solve a first-order differential equation by separating the variables
- solve problems involving exponential growth or decay

A Forming a differential equation (answers p 182)

Imagine a population of organisms (say bacteria) in which individual organisms reproduce. The more organisms there are in the population, the faster the population will grow. Suppose that the rate of growth is proportional to the size of the population.

Let the population at time t hours be P.

The rate of growth of the population, in organisms per hour, is given by $\dfrac{\mathrm{d}P}{\mathrm{d}t}$.

The fact that the rate of growth is proportional to the size of the population can be written as

$$\frac{\mathrm{d}P}{\mathrm{d}t} \propto P \ \text{ or } \ \frac{\mathrm{d}P}{\mathrm{d}t} = kP, \text{ where } k \text{ is a constant.}$$

Suppose that the value of k for a particular population is known to be 0.1.

Then

$$\frac{\mathrm{d}P}{\mathrm{d}t} = 0.1P$$

An equation that involves a derivative (or more than one derivative) is called a **differential equation**. Differential equations are used to model situations involving rates of change.

The differential equation above is a **first-order** differential equation as it involves a first-order derivative $\dfrac{\mathrm{d}P}{\mathrm{d}t}$ but no derivatives of higher order, such as $\dfrac{\mathrm{d}^2P}{\mathrm{d}t^2}$.

A1 The rate of growth of a population P at time t is given by $0.05P^2$.
Write this information as a differential equation.

A2 The temperature of an oven at time t minutes is $T\,°C$. The rate at which the temperature is increasing is given by $(2t + 0.01t^2)$ degrees per minute.
Write this information as a differential equation.

A3 The population P of a region at time t years is **decreasing** at a rate given by $0.1Pt$.
Explain why it is incorrect to write this information as $\dfrac{\mathrm{d}P}{\mathrm{d}t} = 0.1Pt$, and write the differential equation correctly.

A4 An object moves along a straight track. Its distance at time t seconds is s metres. The distance decreases at a rate given by $0.02s$ metres per second.
Write this information as a differential equation.

A5 The volume of water in a tank at time t seconds is $V\,\mathrm{m}^3$.
Water flows into the tank at a rate given by $0.05t^2\,\mathrm{m}^3\,\mathrm{s}^{-1}$.
Water also flows out at constant rate of $4\,\mathrm{m}^3\,\mathrm{s}^{-1}$.

Write this information as a differential equation.

A6 The area of weed on a lake at time t seconds is $A\,\mathrm{m}^3$.
The area of weed grows at a rate that is directly proportional to the area.

Write this information as a differential equation involving a constant of proportionality k.

A7 The depth of water in a reservoir at time t hours is h metres.
The depth **decreases** at a rate that is proportional to the square root of the depth.

Write this information as a differential equation involving a positive constant of proportionality k.

In some situations, two quantities are connected to one another and this leads to a relationship between their rates of change.

For example, the area, $A\,\mathrm{cm}^2$, of a square of side $s\,\mathrm{cm}$ is given by $A = s^2$.

Suppose that s is increasing with time. Its rate of change is $\dfrac{ds}{dt}$.

The area is also increasing. Its rate of change is $\dfrac{dA}{dt}$.

To find the relationship between $\dfrac{dA}{dt}$ and $\dfrac{ds}{dt}$, we differentiate both sides of the equation $A = s^2$ with respect to t.

We use the chain rule to differentiate s^2 with respect to t: $\dfrac{d}{dt}\left(s^2\right) = \dfrac{d}{ds}\left(s^2\right)\dfrac{ds}{dt} = 2s\dfrac{ds}{dt}$

So $\qquad A = s^2$

leads to $\quad \dfrac{dA}{dt} = 2s\dfrac{ds}{dt}$.

A8 The area, $A\,\mathrm{cm}^2$, and the radius, $r\,\mathrm{cm}$, of a circle are connected by the equation $A = \pi r^2$.

Given that r and A are the radius and area at time t seconds, find the equation connecting $\dfrac{dA}{dt}$ and $\dfrac{dr}{dt}$.

A9 A sphere is growing in size. At time t seconds, the radius of the sphere is $r\,\mathrm{cm}$ and the volume is $V\,\mathrm{cm}^3$, where $V = \frac{4}{3}\pi r^3$.

Find the equation connecting $\dfrac{dV}{dt}$ and $\dfrac{dr}{dt}$.

A10 The equation $A = s^2$ for a square can also be written as $s = A^{\frac{1}{2}}$.
Use this form of the equation to find an equation linking $\dfrac{ds}{dt}$ and $\dfrac{dA}{dt}$.

A11 At time t seconds, the side of a square is of length s cm and the area of the square is A cm^2.

The length of the side is increasing at a rate that is proportional to the length.

(a) Write down the differential equation for $\dfrac{ds}{dt}$ in terms of s, using k for the constant of proportionality.

(b) Differentiate both sides of the equation $A = s^2$ to get the equation connecting $\dfrac{dA}{dt}$ and $\dfrac{ds}{dt}$.

(c) Hence show that A satisfies the differential equation $\dfrac{dA}{dt} = 2kA$.

Example 1

At time t seconds, the radius of a sphere is r cm and the surface area of the sphere is S cm^2, where $S = 4\pi r^2$.

(a) Given that the radius is increasing at a rate that is proportional to the square of the radius, show that S satisfies a differential equation of the form $\dfrac{dS}{dt} = aS^{\frac{3}{2}}$, where a is a constant.

(b) Given that the area is increasing at a rate that is inversely proportional to the area, show that r satisfies a differential equation of the form $\dfrac{dr}{dt} = \dfrac{b}{r^3}$, where b is a constant.

Solution

For both parts of the question, first differentiate both sides of $S = 4\pi r^2$ with respect to t.

$$S = 4\pi r^2$$
$$\frac{dS}{dt} = 8\pi r \frac{dr}{dt} \quad (1)$$

(a) *You are given that $\dfrac{dr}{dt} \propto r^2$, or $\dfrac{dr}{dt} = kr^2$, where k is a constant.*

Substitute this into equation (1).

$$\frac{dS}{dt} = 8\pi k r^3 \quad (2)$$

You now need to express r^3 in terms of S, using once again the formula $S = 4\pi r^2$.

$$S = 4\pi r^2 \implies r = \left(\frac{S}{4\pi}\right)^{\frac{1}{2}} \implies r^3 = \left(\frac{S}{4\pi}\right)^{\frac{3}{2}}$$

So from equation (2) it follows that $\dfrac{dS}{dt} = 8\pi k \left(\dfrac{S}{4\pi}\right)^{\frac{3}{2}}$.

Everything on the right-hand side, apart from $S^{\frac{3}{2}}$, is a constant, so the equation may be written as $\dfrac{dS}{dt} = aS^{\frac{3}{2}}$, where a is a constant.

(b) *You are given that* $\dfrac{dS}{dt} \propto \dfrac{1}{S}$ *or* $\dfrac{dS}{dt} = \dfrac{k}{S}$, *where k is a constant.*

Substitute into equation (1). $\quad \dfrac{k}{S} = 8\pi r \dfrac{dr}{dt}$

Use $S = 4\pi r^2$ *to eliminate S.* $\quad \dfrac{k}{4\pi r^2} = 8\pi r \dfrac{dr}{dt}$, *from which* $\dfrac{dr}{dt} = \dfrac{1}{32\pi^2 r^3} = \dfrac{b}{r^3}$

Exercise A (p 182)

1 Water enters a reservoir at a constant rate of 200 litres per second. Water also leaks out of the reservoir at a rate of $0.1V$ litres per second, where V litres is the volume of water at time t seconds.

Show that $-10\dfrac{dV}{dt} = V - 2000$.

2 The intensity of illumination on a small screen placed at a distance from a light source is inversely proportional to the square of the distance.

At time t seconds, the distance of the screen from the source is x cm and the intensity of illumination is I units.

(a) Express the relationship between I and x as an equation involving a constant of proportionality k.

(b) Find the equation connecting $\dfrac{dI}{dt}$ and $\dfrac{dx}{dt}$.

(c) The screen moves away from the source with a speed proportional to its distance from the source. Show that $\dfrac{dI}{dt} = -aI$, where a is a positive constant.

3 A container is in the shape of a horizontal triangular prism whose cross-section is an equilateral triangle, as shown in the diagram.

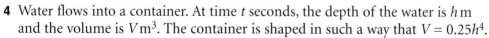

Water flows into the container. At time t seconds, the depth of the water is h m and the volume is V m³.

(a) Explain why $V \propto h^2$.

(b) Given that the water flows in at a constant rate, show that $\dfrac{dh}{dt}$ is inversely proportional to h.

4 Water flows into a container. At time t seconds, the depth of the water is h m and the volume is V m³. The container is shaped in such a way that $V = 0.25h^4$.

(a) Find the equation connecting $\dfrac{dV}{dt}$ and $\dfrac{dh}{dt}$.

(b) Given that the rate at which water flows in is proportional to the square root of the volume of water in the container, show that $\dfrac{dh}{dt}$ is inversely proportional to h.

5 The 'kinetic energy' of a moving object is proportional to the square of its speed. At time t seconds, the speed of an object is v m s⁻¹ and its kinetic energy is E joules.

Given that the speed is increasing at a constant rate, show that $\dfrac{dE}{dt} \propto \sqrt{E}$.

B Solving by separating variables (answers p 183)

Here again is the differential equation for the population of organisms described at the beginning of the previous section:

$$\frac{dP}{dt} = 0.1P$$

Solving this differential equation means finding the equation connecting P and t. The process of solving will involve integration.

It is possible to rewrite this differential equation so that the left-hand side of the equation involves only P and the right-hand side only t. This is called **separating the variables**.

The first step is to separate the numerator and denominator of $\frac{dP}{dt}$.

(You have met this idea when doing integration by substitution.) $\qquad dP = 0.1P\,dt$

Now divide both sides by P (or multiply by $\frac{1}{P}$), so that the variables are completely separated.

$$\frac{1}{P}\,dP = 0.1\,dt$$

With the equation in this form, both sides can be integrated.

$$\int \frac{1}{P}\,dP = \int 0.1\,dt$$

$$\ln|P| = 0.1t + c$$

By 'exponentiating' both sides (that is, $e^{\text{left side}} = e^{\text{right side}}$), we get

$$e^{\ln|P|} = e^{0.1t + c}$$
$$= e^{0.1t}e^{c}$$

c is just a constant whose value at this stage is unknown. So is e^c. We can let A stand for e^c, so that A is now the unknown constant.

$$= Ae^{0.1t}$$

Because $e^{\ln|P|} = |P|$ it follows that

$$|P| = Ae^{0.1t}$$

In this case we know that $P > 0$, so we can write

$$P = Ae^{0.1t}$$

This equation is called the **general solution** of the differential equation.
It contains an unknown constant A.
To fix the value of A we need some more information.
Suppose we know that when $t = 0$, the value of P is 200. Then $\qquad 200 = Ae^0$, so $A = 200$

The resulting equation is called a **particular solution**. $\qquad P = 200e^{0.1t}$

K The method of solving a differential equation by separating the variables can be summarised as follows.

- Rearrange the equation so that the left-hand side involves only one variable and the right-hand side only the other.

- Integrate each side to get the general solution of the differential equation. This will include an unknown constant.

- Use information about known values of the variables to fix the value of the constant. (This information is sometimes called the 'boundary conditions'.) The resulting solution is the particular solution which fits the given information.

B1 By differentiating, show that if $P = Ae^{0.1t}$ then $\frac{dP}{dt} = 0.1P$.

(This verifies that $P = Ae^{0.1t}$ is the general solution of the differential equation.)

B2 **(a)** Given that $\dfrac{dP}{dt} = \dfrac{5}{P}$, $P > 0$,

      (i) complete this rearrangement of the differential equation: $P\,dP = \ldots$

      (ii) by integrating each side, show that $P^2 = 10t + A$, where A is a constant

  (b) Verify by differentiation that if $P = \sqrt{10t + A}$ then $\dfrac{dP}{dt} = \dfrac{5}{P}$.

  (c) Given that $P = 6$ when $t = 0$, show that $A = 36$.

  (d) Find the value of P when $t = 10.8$.

B3 **(a)** Given that $\dfrac{dP}{dt} = 0.01P^2$,

      (i) complete this rearrangement of the differential equation: $\dfrac{1}{P^2}\,dP = \ldots$

      (ii) by integrating each side, show that $P = -\dfrac{1}{0.01t + c}$

  (b) Given that $P = 10$ when $t = 0$, show that $c = -0.1$ and hence that $P = \dfrac{100}{10 - t}$.

  (c) What happens to P as t gets closer and closer to 10?

D **B4** A student was given this differential equation: $\dfrac{dy}{dx} = 3y^2$.

  (a) Explain what is wrong with this working:

$$dy = 3y^2\,dx$$
$$\int dy = \int 3y^2\,dx$$
$$\int 1\,dy = \int 3y^2\,dx$$
$$y = y^3 + c$$

  (b) Show that the correct general solution of the equation is $y = -\dfrac{1}{3x + c}$.

Example 2

Show that the general solution of the differential equation $\dfrac{dy}{dx} = 2x(y + 4)$, $y > 0$,

is $y = Ae^{x^2} - 4$.

Solution

First separate the variables. $\dfrac{dy}{y + 4} = 2x\,dx$ $\left(\dfrac{dy}{y + 4} \text{ is the same as } \dfrac{1}{y + 4}\,dy.\right)$

Integrate both sides. $\displaystyle\int \dfrac{1}{y + 4}\,dy = \int 2x\,dx$

          $\Rightarrow \quad \ln(y + 4) = x^2 + c$ Since $y > 0$, $|y + 4|$ *is unnecessary.*

Exponentiate both sides. $y + 4 = e^{x^2 + c} = e^{x^2}e^c = Ae^{x^2}$ (where $A = e^c$)

          $\Rightarrow \qquad\qquad y = Ae^{x^2} - 4$

Example 3

A water tank is filled in such a way that the rate at which the depth of the water increases is proportional to the square root of the depth.

Initially the depth is 4 m. After a time t hours the depth is h m.

(a) Write down a differential equation for h.

(b) Show that $\sqrt{h} = \frac{1}{2}kt + 2$, where k is a constant.

(c) Given that $h = 16$ when $t = 6$, find the value of k.

(d) Find the time taken to fill the tank to a depth of 36 m.

Solution

(a) The rate of increase of h is $\dfrac{dh}{dt}$, and this is proportional to $\sqrt{h}$. So $\dfrac{dh}{dt} = k\sqrt{h}$.

(b) *Separate the variables and integrate.* $\qquad \dfrac{dh}{\sqrt{h}} = k\,dt$

$$\int h^{-\frac{1}{2}}\,dh = \int k\,dt$$

$$\Rightarrow \qquad 2h^{\frac{1}{2}} = kt + c$$

$$\Rightarrow \qquad \sqrt{h} = \tfrac{1}{2}(kt + c)$$

When $t = 0,\, h = 4$, so $\qquad \sqrt{4} = \tfrac{1}{2}(k \times 0 + c)$

$$\Rightarrow \qquad 2 = \tfrac{1}{2}c \text{ so } c = 4$$

so $\qquad \sqrt{h} = \tfrac{1}{2}(kt + 4) = \tfrac{1}{2}kt + 2$

(c) $4 = 3k + 2$, so $k = \tfrac{2}{3}$.

(d) The equation for $\sqrt{h}$ is $\sqrt{h} = \tfrac{1}{3}t + 2$. When $h = 36$, $6 = \tfrac{1}{3}t + 2$, so $t = 12$.

Exercise B (answers p 184)

1 (a) Show that the general solution of the differential equation $\dfrac{dy}{dx} = \dfrac{x^3}{y}$ can be written in the form $y^2 = \tfrac{1}{2}x^4 + A$, where A is a constant.

   (b) Find the particular solution for which $y = 4$ when $x = 0$.

2 (a) Show that the general solution of the differential equation $\dfrac{dy}{dx} = \dfrac{1+x}{y}$ can be written in the form $y^2 = x^2 + 2x + A$, where A is a constant.

   (b) Find the value of A given that $y = 3$ when $x = 1$.

3 (a) Show that the general solution of the differential equation $\dfrac{dy}{dx} = xy$ can be written in the form $y = Ae^{\frac{1}{2}x^2}$, where A is a constant.

   (b) Find the value of A given that $y = 6$ when $x = 0$.

4 Solve the differential equation $\dfrac{dy}{dx} = 1 + y$ $(y > 0)$ given that $y = 5$ when $x = 0$.

5 **(a)** Show that the general solution of the differential equation $\dfrac{dy}{dx} = 2x\sqrt{y}$

can be written in the form $\sqrt{y} = \frac{1}{2}x^2 + A$, where A is a constant.

(b) Given that $y = 4$ when $x = 0$, find the value of A.

(c) Find, to three significant figures, the value of x for which $y = 20$.

6 As a result of a disease, the population of a colony of animals is decreasing at a rate proportional to the size of the population. Initially the population is $20\,000$; after time t days the population is P.

(a) Write down a differential equation for P, including a constant of proportionality k.

(b) Show that the general solution of the differential equation can be written in the form $P = Ae^{-kt}$, where A is a constant.

(c) Use the fact that initially $P = 20\,000$ to find the value of A.

(d) Given that $P = 10\,000$ when $t = 10$, find the value of k to three significant figures.

(e) Find, to three significant figures, the value of t for which $P = 1000$.

7 **(a)** Find the general solution of the differential equation $\dfrac{dy}{dx} = \dfrac{y}{x^2}$ $(x > 0, y > 0)$, giving y in terms of x.

(b) Find the particular solution for which $y = e$ when $x = 1$.

8 Solve the differential equation $\dfrac{dy}{dx} = xy^2$ given that $y = 1$ when $x = 0$.

9 Solve the differential equation $\dfrac{dy}{dx} = \dfrac{y^2}{x}$ $(x > 0, y > 0)$ given that $y = \frac{1}{2}$ when $x = 1$.

10 **(a)** Find $\displaystyle\int \dfrac{x}{x^2+1}\,dx$.

(b) **(i)** Show that the general solution of the differential equation $\dfrac{dy}{dx} = \dfrac{xy}{x^2+1}$,

where $x \geq 0$ and $y > 0$, can be expressed in the form $y = A\sqrt{x^2+1}$, where A is a constant.

(ii) Given that $y = 10$ when $x = 1$, find the exact value of A.

(c) **(i)** Find the general solution of the differential equation $\dfrac{dy}{dx} = \dfrac{xy^2}{x^2+1}$ $(x \geq 0, y > 0)$.

(ii) Given that $y = 1$ when $x = 0$, show that $y = \dfrac{1}{1 - \frac{1}{2}\ln(x^2+1)}$.

11 Newton's law of cooling states that the rate at which an object's temperature decreases is proportional to the difference between the object's temperature and the temperature of its surroundings.

(a) Write this as a differential equation using $T\,°C$ for the object's temperature and $A\,°C$ for the surrounding temperature.

(b) Show that the general solution can be written as $T = A + Be^{-kt}$, where k and B are constants.

C Exponential growth and decay (answers p 185)

This is the differential equation for the population of organisms described at the beginning of section A:

$$\frac{dP}{dt} = 0.1P$$

The rate of growth of the population is proportional to the size of the population. The constant of proportionality is 0.1.

In section B it was shown that the general solution of the equation is $P = Ae^{0.1t}$

For the case where $P = 200$ at $t = 0$, the particular solution is $P = 200e^{0.1t}$

The graph of P against t is shown here. (Both negative and positive values of t are shown.)

This is an example of **exponential growth**.

The general form of an exponential growth function of t is Ae^{bt}, where A and b are both positive.

If $P = Ae^{bt}$, then $\frac{dP}{dt} = Abe^{bt} = bP$, which shows that the rate of growth is proportional to the value of P.

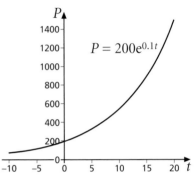

C1 Given that $P = 200e^{0.1t}$, find, to three significant figures, the value of t for which P is

(a) 300   (b) 400   (c) 150

If A is positive but b is negative in the equation $P = Ae^{bt}$, then we have **exponential decay**.

For example, here is the graph of $P = 50e^{-0.5t}$.

C2 (a) Given that $P = 50e^{-0.5t}$, find, to three significant figures, the value of t for which P is

(i) 30   (ii) 20   (iii) 60

(b) What happens to P as t gets larger and larger?

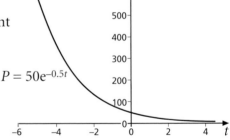

Exponential decay functions are often used to model processes where a quantity gets closer and closer to a given limiting value as time goes on.

The phrase 't gets larger and larger' is usually expressed in symbols as '$t \to \infty$' (read as 't tends to infinity').

As $t \to \infty$, the value of Ae^{-bt}, where $b > 0$, gets closer and closer to 0, so the value of $C + Ae^{-bt}$ gets closer and closer to C from above.

The value of $C - Ae^{-bt}$ gets closer and closer to C from below.

This is illustrated in the graphs on the right.

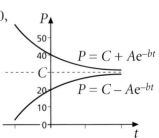

C3 Given that $y = 40 + 10e^{-2t}$,

    (a) find the value of y when $t = 0$

    (b) find the limiting value of y as $t \to \infty$

    (c) sketch the graph of y against t

C4 A sample of liquid is heated to a temperature of 80 °C and then allowed to cool down. The temperature gradually falls towards that of the surrounding atmosphere, which is 20 °C. The equation $T = c + ae^{-bt}$ $(b > 0)$ is used to model the cooling process, where T °C is the temperature of the liquid and t is the time in minutes since cooling started.

    (a) By considering what happens as $t \to \infty$, find the value of c.

    (b) By considering the situation at $t = 0$, find the value of a.

    (c) Given that it takes 10 minutes for the temperature to fall to 35 °C, show that $e^{-10b} = 0.25$ and hence find the value of b to three significant figures.

    (d) Find the rate of cooling, in degrees per minute, when the temperature of the liquid is 35 °C.

> An equation of the form $y = ae^{bt}$ $(a > 0, b > 0)$ represents exponential growth.
>
> An equation of the form $y = ae^{-bt}$ $(a > 0, b > 0)$ represents exponential decay.
>
> The equation $y = c \pm ae^{-bt}$ $(a > 0, b > 0)$ represents a process in which the value of y gets closer and closer to c as $t \to \infty$.

Example 4

A population P is modelled by the equation $P = r - se^{-0.5t}$, where r and s are constants and t is the time in years.

(a) Given that $P = 200$ when $t = 0$ and $P = 360$ when $t = 4$, find the values of r and s to three significant figures.

(b) What happens to the population as t gets larger and larger?

(c) At what rate is the population growing when $t = 6$?

Solution

(a) *Substitute the given values into the equation.*
$$200 = r - s$$
$$360 = r - se^{-2}$$

Subtract the first equation from the second.
$$160 = s(1 - e^{-2})$$

$$\Rightarrow \quad s = \frac{160}{1 - e^{-2}} = 185 \text{ to 3 s.f.}$$

Use the first equation to find the value of r.
$$r = 200 + 185 = 385 \text{ to 3 s.f.}$$

(b) The equation for P is $P = 385 - 185e^{-0.5t}$.
As $t \to \infty$, P gets closer and closer to 385.

(c) When $t = 6$, $\dfrac{dP}{dt} = -185 \times (-0.5e^{-0.5t}) = 92.5e^{-3} = 4.61$ per year (to 3 s.f.)

Exercise C (answers p 185)

1 The growth of a population P is modelled by the equation $P = 5000e^{0.04t}$, where t is the time in years from the first count.

  (a) What was the size of the population when it was first counted?

  (b) What is the population at time $t = 20$?

  (c) Find the rate of increase at time $t = 20$, to the nearest whole number.

2 Given that $y = 50 - 10e^{-\frac{1}{2}t}$, find

  (a) the value of y when $t = 0$ **(b)** the limiting value of y as $t \to \infty$

3 The size of a population is modelled by the equation $P = ae^{-kt}$.

  (a) Given that $P = 800$ when $t = 0$, write down the value of a.

  (b) Given that $P = 500$ when $t = 5$, show that $k = 0.0940$ to 3 s.f.

  (c) Find, to three significant figures,

   (i) the value of P when $t = 4$ **(ii)** the value of t for which $P = 400$

4 The temperature, $T\,°C$, at the centre of a bonfire is modelled by the equation $T = 200 - qe^{-kt}$, where t is the time in minutes since the fire was lit and $k > 0$.

  (a) The temperature at the centre when the fire was lit was $15\,°C$. Find the value of q.

  (b) What happens to the temperature as $t \to \infty$?

  (c) It takes 10 minutes for the centre to reach a temperature of $100\,°C$. Find the value of k, to three significant figures.

5 When a radioactive isotope decays, the number of atoms that remain radioactive after t years is given by $N = N_0e^{-kt}$, where N_0 is the initial number of radioactive atoms and k is a constant. In a particular case $k = 0.04$. Show that the time taken for half the atoms to lose their radioactivity is about 17.3 years. (This called the 'half-life' of the substance.)

D Further exponential functions (answers p 185)

Exponential functions are normally expressed in terms of e, because the derivative of e^x or e^t is simple to find and because logarithms to base e are readily available on calculators and as a spreadsheet function.

However, for any positive value of a, the functions a^x and a^{-x} are exponential functions too and they could be used to model exponential growth or decay.

There is a straightforward way of converting an exponential function of the form a^x into the normal exponential form that uses e.

D1 **(a)** Use the fact that $5 = e^{\ln 5}$ to write 5^x in the form e^{kx}, giving k to 3 s.f.

  (b) Use the fact that $8 = e^{\ln 8}$ to write 8^x in the form e^{kx}.

  (c) Write 2^x in the form e^{kx}.

  (d) Write a^x in the form e^{kx}.

(K) The expression a^x is equivalent to $e^{(\ln a)x}$.

D2 Show that the derivative of a^x is $(\ln a)a^x$.

(K) The derivative of a^x is $(\ln a)a^x$.

D3 The value £V of a car t years after initial purchase is modelled by the equation $V = 20\,000 \times 0.85^t$.

   (a) What was the initial value of the car?

   (b) What is its value after 5 years?

   (c) The value of t for which $V = 8000$ is given by $20\,000 \times 0.85^t = 8000$, from which $0.85^t = \frac{8000}{20\,000} = 0.4$.

   Use logarithms to find the value of t, to three significant figures.

   (d) Find the value of $\dfrac{dV}{dt}$ when $t = 5$.

D4 The population of an animal colony increases by 20% each year.

   (a) By what number is the population multiplied each year?

   (b) Given that the initial population is 8000, write down an expression for the population after t years.

D5 Redo question D4 for the case where the population decreases by 20% each year.

Example 5

The population P of an animal colony is modelled by the equation $P = A \times (1.08)^t$, where A is a constant and t is the time in years since the population was first counted.

(a) Given that the population 3 years after the first count is 5400, find the population at the first count, to three significant figures.

(b) How many years after the first count will the population reach $10\,000$?

(c) Find the rate at which the population is increasing 3 years after the first count.

Solution

(a) The population at the first count is the value of A.

Since $5400 = A \times (1.08)^3$, it follows that $A = \dfrac{5400}{(1.08)^3} = 4290$ to 3 s.f.

(b) $\qquad 10\,000 = 4290 \times 1.08^t$

$\Rightarrow \quad 1.08^t = \frac{10\,000}{4290}$

$\Rightarrow t \ln 1.08 = \ln\left(\frac{10\,000}{4290}\right)$

$\Rightarrow \qquad t = \dfrac{\ln\left(\frac{10\,000}{4290}\right)}{\ln 1.08} = 11.0$ to 3 s.f.

(c) $\qquad P = 4290 \times 1.08^t$

So $\dfrac{dP}{dt} = 4290 \times \ln 1.08 \times 1.08^t$

When $t = 3$, $\dfrac{dP}{dt} = 4290 \times \ln 1.08 \times 1.08^3$

$= 416$ individuals per year to 3 s.f.

Exercise D (answers p 185)

1 The population P of a city region is modelled by the equation $P = P_0 \times (1.04)^t$, where P_0 is the population now and t is the time in years measured from now. Given that $P_0 = 240\,000$, find

(a) the time taken from now for the population to reach $300\,000$

(b) the rate of increase of the population 4 years from now

2 The value £V of a vehicle is modelled by the equation $V = 16\,000 \times (0.8)^t$, where t is the time in years since the vehicle was purchased. Find

(a) the time taken from the time of purchase for the vehicle to halve in value

(b) the rate at which the value is decreasing 1 year after purchase

3 Given that $y = p + qa^x$, where p, q and a are constants, show that $\dfrac{dy}{dx} = (\ln a)(y - p)$.

Key points

- To solve a differential equation by separating variables, rewrite the equation so that each variable appears on only one side, integrate each side to find the general solution and use any additional information (boundary conditions) to find the particular solution. (p 100)

- An equation of the form $y = ae^{bt}$ $(a > 0, b > 0)$ represents exponential growth. An equation of the form $y = ae^{-bt}$ $(a > 0, b > 0)$ represents exponential decay. The equation $y = c \pm ae^{-bt}$ $(a > 0, b > 0)$ represents a process in which the value of y gets closer and closer to c as $t \to \infty$. (p 105)

- The expression a^x is equivalent to $e^{(\ln a)x}$. The derivative of a^x is $(\ln a)a^x$. (p 107)

Mixed questions (answers p 185)

1 A researcher is modelling the sales of a music CD. She suggests that if the total number sold t weeks from launching the CD is N, then the rate at which the total grows is proportional to $N(20\,000 - N)$.

(a) Write the suggested model as a differential equation, including a constant k.

(b) Explain why the model predicts that the total number sold will not exceed $20\,000$.

2 The volume, $V\,\text{cm}^3$, of a sphere of radius $r\,\text{cm}$ is given by $V = \frac{4}{3}\pi r^3$.

(a) Given that the radius r at time $t\,$s is increasing at a rate that is proportional to r, show that $\dfrac{dV}{dt} = aV$, where a is a constant.

(b) Given that the volume V at time $t\,$s is decreasing at a rate that is proportional to V, show that $\dfrac{dr}{dt} = -br$, where b is a positive constant.

3 Solve the differential equation

$$\frac{dy}{dx} = \frac{1+y}{1+x}$$

given that $y = 0$ when $x = 0$.

4 Liquid is poured into a reservoir whose shape is an upturned pyramid.
At time t minutes the depth of the water is h m.
The rate at which h increases is inversely proportional to h^2.

(a) Write down a differential equation for h, including a constant of proportionality.

(b) Show that the general solution of the differential equation can be written in the form $h^3 = At + B$, where A and B are constants.

(c) Given that $h = 4$ when $t = 0$ and $h = 5$ when $t = 1$, find the values of A and B.

5 The surface area, S cm^2, of a sphere of radius r cm is given by $S = 4\pi r^2$.

(a) At time t s, the radius r is increasing at a rate given by $\frac{1}{4}r$ cm s^{-1}.

(i) Express $\frac{dS}{dt}$ in terms of $\frac{dr}{dt}$. (ii) Show that $\frac{dS}{dt} = \frac{1}{2}S$.

(b) Given that $S = 5$ when $t = 0$, find S in terms of t.

6 (a) Find constants A and B such that $\dfrac{x+1}{(x-2)(x-1)} \equiv \dfrac{A}{x-2} + \dfrac{B}{x-1}$.

(b) Solve the differential equation $\dfrac{dy}{dx} = \dfrac{x+1}{(x-2)(x-1)}$ $(x \geq 3)$
given that $y = 0$ when $x = 3$.

7 Water flows into a vertical tank that is in the shape of a cylinder.
At time t seconds the depth of water in the tank is h m and the volume V m^3.
The rate at which the water flows in is inversely proportional to V.

(a) Show that $\dfrac{dh}{dt} = \dfrac{k}{h}$, where k is a constant.

(b) Show that the general solution of this equation, for $h > 0$, may be written as $h = \sqrt{At + B}$, where A and B are constants.

At time $t = 0$, the depth of the water in the tank is 2 m.
5 seconds later, the depth has increased to 5 m.

(c) Find the values of A and B.

8 Solve the differential equation

$$\frac{dy}{dx} = \sqrt{xy} \quad x \geq 0, \ y \geq 0$$

given that $y = 1$ when $x = 0$.

9 Solve the differential equation

$$\frac{\mathrm{d}y}{\mathrm{d}x} = (x+1)(y+1)$$

given that $y = 0$ when $x = 0$.

10 (a) Find $\displaystyle\int \frac{y}{1+y^2}\,\mathrm{d}y$.

(b) Solve the differential equation

$$y\frac{\mathrm{d}y}{\mathrm{d}x} = 1 + y^2$$

given that $y = 0$ when $x = 0$.

11 Show that the general solution of the differential equation

$$\frac{\mathrm{d}y}{\mathrm{d}x} = \cos^2 x \cos^2 y$$

may be written as $y = \arctan\left(\frac{1}{4}\sin 2x + x + c\right)$, where c is a constant.

12 A wire is stretched from a fixed point A.
The wire passes over a movable bridge B.

The frequency, f hertz, of the note emitted when the part AB of the wire vibrates is inversely proportional to the length L m of AB.

While the wire is vibrating, L is increased at a rate that is proportional to $\sqrt{L}$.

Given that f is the frequency at time t seconds,

(a) show that f satisfies the differential equation

$$\frac{\mathrm{d}f}{\mathrm{d}t} = -kf^{\frac{3}{2}}$$

where k is a positive constant

(b) show that the general solution of this equation may be written as

$$f = \frac{4}{(kt+c)^2}$$

where c is a constant

13 The rotational speed, r revolutions per second, of a flywheel satisfies the differential equation $\dfrac{\mathrm{d}r}{\mathrm{d}t} = \dfrac{50-r}{2}$.

(a) Find the general solution of this equation and show that it can be written in the form $r = 50 - A\mathrm{e}^{-\frac{1}{2}t}$.

(b) What happens to r as $t \to \infty$?

(c) Given that $r = 20$ when $t = 0$, find the value of A.

(d) Find, to three significant figures, the value of t for which $r = 40$.

14 (a) Express $\dfrac{1}{y^2-1}$ as the sum of partial fractions.

(b) Hence show that the general solution of the differential equation
$$\frac{dy}{dx} = y^2 - 1 \quad (y > 1)$$
may be written in the form $y = \dfrac{1 + Ae^{2x}}{1 - Ae^{2x}}$, where A is a constant.

Test yourself (answers p 187)

1 (a) Solve the differential equation $\dfrac{dy}{dx} = \dfrac{y-3}{2}$, giving the general solution for y in terms of x.

(b) Find the particular solution of this differential equation for which $y = 2$ when $x = 0$.

2 Show that the general solution of the differential equation $\dfrac{dy}{dx} = x(y+1)$ $(y > 0)$ can be written in the form $y = Ae^{\frac{1}{2}x^2} - 1$.

3 (a) Express $\dfrac{13-2x}{(2x-3)(x+1)}$ in partial fractions.

(b) Given that $y = 4$ at $x = 2$, use your answer to part (a) to find the solution of the differential equation
$$\frac{dy}{dx} = \frac{y(13-2x)}{(2x-3)(x+1)}, \quad x > 1.5$$
Express your answer in the form $y = f(x)$.

Edexcel

4 A circular stain grows in such a way that the rate of increase of its radius is inversely proportional to the square of the radius.
Given that the area of the stain at time t seconds is $A \, \text{cm}^2$,

(a) show that $\dfrac{dA}{dt} \propto \dfrac{1}{\sqrt{A}}$.

Another stain, which is growing more quickly, has area $S \, \text{cm}^2$ at time t seconds.
It is given that $\dfrac{dS}{dt} = \dfrac{2e^{2t}}{\sqrt{S}}$.
Given that, for this second stain, $S = 9$ at time $t = 0$,

(b) solve the differential equation to find the time at which $S = 16$.
Give your answer to two significant figures.

Edexcel

9 Vectors

In this chapter you will work in two and three dimensions.
You will learn how to

- find the magnitude of a vector
- add and subtract vectors and their scalar multiples
- use column vectors and **i**, **j**, **k** notation
- work with position vectors
- find and use the vector equation of a straight line
- find the scalar product of two vectors
- use the scalar product to find the angle between two straight lines

A Vectors in two dimensions (answers p 188)

For those who have studied Mechanics 1 this section is mainly revision.

A1 Towns A and B are 40 km apart and towns B and C are 30 km apart.
Is this enough information to find the distance between towns A and C?

This sketch shows the relative positions of towns A, B and C.

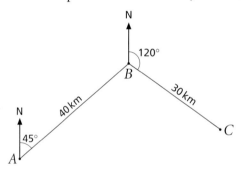

A2 (a) Find the size of angle ABC.

(b) Calculate the distance between towns A and C.

The **distances** AB and BC do not add to give the distance AC.

The **displacement** from A to B (written as $\overrightarrow{AB}$) has a magnitude (which is the distance of 40 km) and a direction (NE).

The displacement from A to B followed by the displacement from B to C is equivalent to the displacement from A to C; we can write this as

$$\overrightarrow{AB} + \overrightarrow{BC} = \overrightarrow{AC}$$

A quantity which behaves in this way is called a **vector** quantity.
A vector quantity has both **magnitude** and **direction**.

A quantity such as temperature or mass that has magnitude but no direction is called a **scalar** quantity.

We can represent a vector by a straight line segment where the length of the line represents the magnitude of the vector and the direction of the line (indicated by an arrow) represents the direction of the vector.

For example, using a scale of 1 cm to 10 km, we can represent the vector $\overrightarrow{AB}$ by this line.

The magnitude of a vector is sometimes called the **modulus**.
The modulus of the vector $\overrightarrow{AB}$ is written as $\left|\overrightarrow{AB}\right|$.

Here we have $\left|\overrightarrow{AB}\right| = 40$.

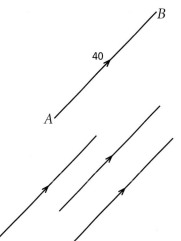

Two vectors are **equal** if they have the same magnitude and the same direction.
So any of these lines may represent the vector $\overrightarrow{AB}$.

An alternative way to label a vector is to use a lower-case letter which

- is in bold type in printed material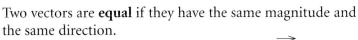

- has a wavy or straight line underneath when handwritten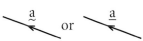

All vectors are added in the same way as the displacements on the previous page. This is sometimes called the 'triangle law'.

To add these two vectors …   … put them 'head to tail' …

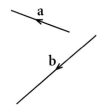 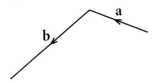 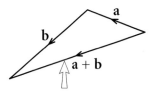

… and this vector represents the vector sum or **resultant**.

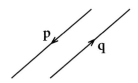 If two vectors **p** and **q** have the same magnitude but opposite directions then we say that **q** = −**p**.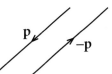

The resultant vector **a** – **b** is equivalent to **a** + –**b**.

To find **a** – **b** ...   ... add **a** and –**b** ...   ... to obtain the resultant **a** + –**b**.

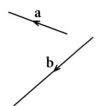   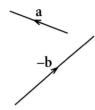   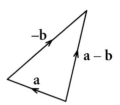

A vector can be multiplied by a number.

For example

- 2**x** is a vector with a magnitude that is twice the magnitude of **x** and in the same direction

- $-\frac{1}{3}$**x** is a vector with a magnitude a third of the magnitude of **x** and in the opposite direction

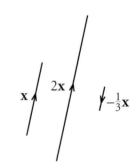

Clearly the vectors **x**, 2**x** and $-\frac{1}{3}$**x** are parallel.
Parallel vectors are sometimes called **scalar multiples** of each other.

K   In general, where k is a number, vectors **a** and k**a** are parallel.

A3 Vectors **a**, **b** and **c** are shown below on square centimetre dotty paper.

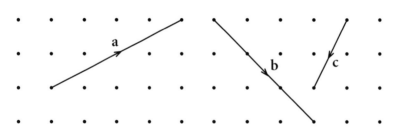

Draw diagrams to show the following vectors.

(a) 2**c**   (b) $\frac{1}{2}$**a**   (c) $-\frac{1}{3}$**b**   (d) **a** + **b**

(e) **a** – **b**   (f) **b** – **a**   (g) $\frac{1}{2}$**a** + 3**b**   (h) **a** + **b** + **c**

D   **A4** For any two vectors **x** and **y**, show that **x** + **y** = **y** + **x**.

A5 Write each of these in its simplest form.

(a) **a** + **b** + 4**b**   (b) 2**a** + **b** + 3**a** – 5**b**

A6 Show that for any two vectors **p** and **q**, the vector **p** + 2**q** is parallel to 2**p** + 4**q**.

A7 Which of the vectors below are parallel to 2**x** – **y**?

    A 2**x** – 3**y**   B 6**x** – 3**y**   C 4**x** + 2**y**   D **y** – 2**x**   E **x** – $\frac{1}{2}$**y**

Example 1

PQRS is a parallelogram.
X is a point on *SR* such that $SX = \frac{1}{3}SR$.

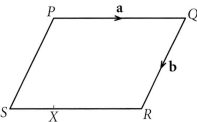

Given that $\overrightarrow{PQ} = \mathbf{a}$ and $\overrightarrow{QR} = \mathbf{b}$, find each of these in terms of $\mathbf{a}$ and $\mathbf{b}$.

(a) $\overrightarrow{SR}$ (b) $\overrightarrow{SX}$ (c) $\overrightarrow{PR}$ (d) $\overrightarrow{QS}$ (e) $\overrightarrow{PX}$ (f) $\overrightarrow{QX}$

Solution

(a) $\overrightarrow{SR} = \overrightarrow{PQ} = \mathbf{a}$
(b) $\overrightarrow{SX} = \frac{1}{3}\overrightarrow{SR} = \frac{1}{3}\mathbf{a}$
(c) $\overrightarrow{PR} = \overrightarrow{PQ} + \overrightarrow{QR} = \mathbf{a} + \mathbf{b}$
(d) $\overrightarrow{QS} = \overrightarrow{QR} + \overrightarrow{RS} = \mathbf{b} + -\mathbf{a} = \mathbf{b} - \mathbf{a}$
(e) $\overrightarrow{PX} = \overrightarrow{PS} + \overrightarrow{SX} = \mathbf{b} + \frac{1}{3}\mathbf{a}$
(f) $\overrightarrow{QX} = \overrightarrow{QR} + \overrightarrow{RX} = \overrightarrow{QR} + \frac{2}{3}\overrightarrow{RS} = \mathbf{b} + \frac{2}{3}(-\mathbf{a}) = \mathbf{b} - \frac{2}{3}\mathbf{a}$

Exercise A (answers p 189)

1 Write each of these in its simplest form.

(a) $\mathbf{x} + \mathbf{y} + 2\mathbf{x}$ (b) $\mathbf{x} + \mathbf{y} - 2\mathbf{x} + 3\mathbf{y}$ (c) $\mathbf{x} - \frac{1}{2}\mathbf{y} + \mathbf{y} - \mathbf{x}$

2 Which of the vectors below are parallel to $\mathbf{p} - 3\mathbf{q}$?

A $2\mathbf{p} - 6\mathbf{q}$ B $4\mathbf{p} - 8\mathbf{q}$ C $\mathbf{q} - \mathbf{p}$ D $\mathbf{q} - \frac{1}{3}\mathbf{p}$ E $4\mathbf{p} + 12\mathbf{q}$

3 *PQRST* is a pentagon.
$\overrightarrow{PQ} = \mathbf{a}$, $\overrightarrow{QR} = \mathbf{b}$, $\overrightarrow{RS} = \mathbf{c}$ and $\overrightarrow{TS} = 3\mathbf{b}$.

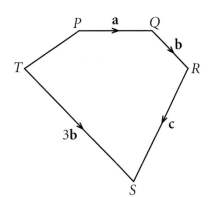

Find each of these in terms of $\mathbf{a}$, $\mathbf{b}$ and $\mathbf{c}$.

(a) $\overrightarrow{PR}$ (b) $\overrightarrow{PS}$ (c) $\overrightarrow{SR}$ (d) $\overrightarrow{TR}$ (e) $\overrightarrow{PT}$

4 *ABCD* is a trapezium.
AB is parallel to *DC* and *DC* = 2*AB*.
M is the mid-point of *DC*.
$\overrightarrow{AB} = \mathbf{p}$ and $\overrightarrow{BC} = \mathbf{q}$.

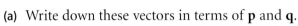

  (a) Write down these vectors in terms of **p** and **q**.

    (i) $\overrightarrow{DC}$ **(ii)** $\overrightarrow{MC}$ **(iii)** $\overrightarrow{BM}$ **(iv)** $\overrightarrow{AD}$

  (b) Show that *ABMD* is a parallelogram.

5 *PQRS* is a quadrilateral.
W, *X*, *Y* and *Z* are the mid-points
respectively of *PQ*, *QR*, *RS* and *PS*.
$\overrightarrow{PQ} = \mathbf{a}$, $\overrightarrow{QR} = \mathbf{b}$, $\overrightarrow{RS} = \mathbf{c}$ and $\overrightarrow{SP} = \mathbf{d}$.

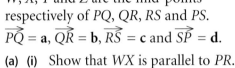

  (a) **(i)** Show that *WX* is parallel to *PR*.

     (ii) Show that *ZY* is parallel to *PR*.

     (iii) Hence, show that *WX* is parallel to *ZY*.

  (b) Show that *XY* is parallel to *WZ*.

  (c) What have you proved about the quadrilateral *WXYZ*?

B Components in two dimensions (answers p 189)

For those who have studied Mechanics 1 this section is mainly revision.

A vector can be written in column vector form.
For example, the vector on the right goes
4 units 'across to the right' and 1 unit 'down'
and, in column vector form, it is written as $\mathbf{a} = \begin{bmatrix} 4 \\ -1 \end{bmatrix}$

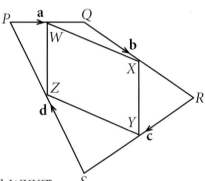

This is the **x-component**.

This is the **y-component**.

Vectors parallel to **a** are scalar multiples of **a**.

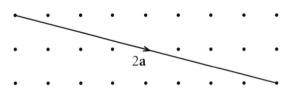

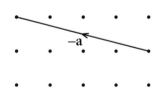

$$2\mathbf{a} = 2\begin{bmatrix} 4 \\ -1 \end{bmatrix} = \begin{bmatrix} 2 \times 4 \\ 2 \times -1 \end{bmatrix} = \begin{bmatrix} 8 \\ -2 \end{bmatrix}$$

$$-\mathbf{a} = -\begin{bmatrix} 4 \\ -1 \end{bmatrix} = \begin{bmatrix} -4 \\ 1 \end{bmatrix}$$

B1 Which of these vectors are parallel to $\begin{bmatrix} 4 \\ -2 \end{bmatrix}$?

  A $\begin{bmatrix} 12 \\ -3 \end{bmatrix}$ **B** $\begin{bmatrix} 8 \\ -4 \end{bmatrix}$ **C** $\begin{bmatrix} 2 \\ -1 \end{bmatrix}$ **D** $\begin{bmatrix} -16 \\ 8 \end{bmatrix}$ **E** $\begin{bmatrix} 20 \\ 10 \end{bmatrix}$

The notation can be used to find resultant vectors when adding or subtracting.

For example, we can see from the diagram that

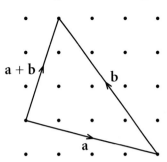

$$\mathbf{a} = \begin{bmatrix} 4 \\ -1 \end{bmatrix}, \ \mathbf{b} = \begin{bmatrix} -3 \\ 4 \end{bmatrix} \text{ and } \mathbf{a} + \mathbf{b} = \begin{bmatrix} 1 \\ 3 \end{bmatrix}$$

Clearly, the resultant vector can be found by adding the x- and y-components:

$$\mathbf{a} + \mathbf{b} = \begin{bmatrix} 4 \\ -1 \end{bmatrix} + \begin{bmatrix} -3 \\ 4 \end{bmatrix} = \begin{bmatrix} 4 + -3 \\ -1 + 4 \end{bmatrix} = \begin{bmatrix} 1 \\ 3 \end{bmatrix}$$

B2 Given that $\mathbf{c} = \begin{bmatrix} 4 \\ -1 \end{bmatrix}$ and $\mathbf{d} = \begin{bmatrix} 2 \\ -1 \end{bmatrix}$, write $\mathbf{c} + 5\mathbf{d}$ as a column vector.

Subtraction works in a similar way. For example, from the diagram,

$$\mathbf{p} = \begin{bmatrix} 2 \\ 1 \end{bmatrix}, \ \mathbf{q} = \begin{bmatrix} -3 \\ 2 \end{bmatrix}$$

and $\mathbf{p} - \mathbf{q} = \mathbf{p} + -\mathbf{q} = \begin{bmatrix} 5 \\ -1 \end{bmatrix}$

Looking at components, $\mathbf{p} - \mathbf{q} = \begin{bmatrix} 2 \\ 1 \end{bmatrix} - \begin{bmatrix} -3 \\ 2 \end{bmatrix} = \begin{bmatrix} 2 - -3 \\ 1 - 2 \end{bmatrix} = \begin{bmatrix} 5 \\ -1 \end{bmatrix}$

We can calculate the magnitude of the vector $\mathbf{p} = \begin{bmatrix} 4 \\ 1 \end{bmatrix}$ using Pythagoras's theorem.

$$|\mathbf{p}| = \sqrt{4^2 + 1^2} = \sqrt{17}$$

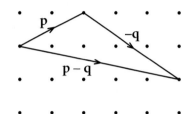

This process will give the magnitude of any vector, including a vector with one or more negative components.

For example, for $\mathbf{q} = \begin{bmatrix} -4 \\ 1 \end{bmatrix}$ we have $|\mathbf{q}| = \sqrt{(-4)^2 + 1^2} = \sqrt{17}$.

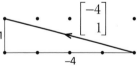

 In general, the magnitude of the vector $\begin{bmatrix} a \\ b \end{bmatrix}$ is $\sqrt{a^2 + b^2}$.

B3 Given that $\mathbf{p} = \begin{bmatrix} 3 \\ -4 \end{bmatrix}$ and $\mathbf{q} = \begin{bmatrix} -5 \\ 12 \end{bmatrix}$, calculate $|\mathbf{p}|$, $|\mathbf{q}|$ and $|\mathbf{p} + \mathbf{q}|$.

Use your results to verify that $|\mathbf{p}| + |\mathbf{q}| \neq |\mathbf{p} + \mathbf{q}|$ in general.

Vectors such as $2\mathbf{p} + 3\mathbf{q}$ and $\mathbf{p} - \mathbf{q}$ are called **linear combinations** of vectors $\mathbf{p}$ and $\mathbf{q}$: they are of the form $a\mathbf{p} + b\mathbf{q}$, where a and b are numbers.

B4 (a) Given that $\mathbf{x} = \begin{bmatrix} 4 \\ -1 \end{bmatrix}$ and $\mathbf{y} = \begin{bmatrix} 3 \\ -4 \end{bmatrix}$, write the linear combination $2\mathbf{x} - \mathbf{y}$ as a column vector.

(b) Find a linear combination of $\mathbf{x}$ and $\mathbf{y}$ that gives the vector $\begin{bmatrix} 13 \\ -13 \end{bmatrix}$.

K Any vector with a magnitude of 1 is a **unit vector**. The unit vectors parallel to the x- and y-axis are conventionally labelled $\mathbf{i}$ and $\mathbf{j}$.

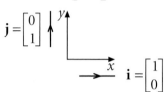

Any vector can be written as a unique linear combination of $\mathbf{i}$ and $\mathbf{j}$.

For example, $\begin{bmatrix} 3 \\ -2 \end{bmatrix} = 3\begin{bmatrix} 1 \\ 0 \end{bmatrix} - 2\begin{bmatrix} 0 \\ 1 \end{bmatrix} = 3\mathbf{i} - 2\mathbf{j}$.

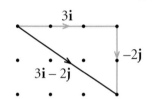

B5 Write $\begin{bmatrix} -4 \\ 3 \end{bmatrix}$ in terms of $\mathbf{i}$ and $\mathbf{j}$.

B6 Write $4\mathbf{i} + 7\mathbf{j}$ as a column vector.

B7 Points P and Q have coordinates $(3, -1)$ and $(5, 4)$ respectively.
(a) Write $\overrightarrow{PQ}$ in terms of $\mathbf{i}$ and $\mathbf{j}$.
(b) Find the distance between points P and Q.

B8 Given that $\mathbf{a} = 2\mathbf{i} + 5\mathbf{j}$ and $\mathbf{b} = \mathbf{i} - 3\mathbf{j}$, write these in terms of $\mathbf{i}$ and $\mathbf{j}$.
(a) $\mathbf{a} + \mathbf{b}$ (b) $\mathbf{a} - \mathbf{b}$ (c) $3\mathbf{a} + 5\mathbf{b}$

B9 What is the magnitude of the vector $8\mathbf{i} - 6\mathbf{j}$?

B10 Write down any vector parallel to $2\mathbf{i} + 5\mathbf{j}$.

B11 (a) Calculate $|10\mathbf{i} + 24\mathbf{j}|$.
(b) Hence write down the magnitude of these vectors.
     (i) $5\mathbf{i} + 12\mathbf{j}$ (ii) $\frac{1}{13}(10\mathbf{i} + 24\mathbf{j})$ (iii) $\frac{10}{3}\mathbf{i} + 8\mathbf{j}$

B12 (a) Calculate the magnitude of $9\mathbf{i} + 12\mathbf{j}$.
(b) Hence write down a vector with a magnitude of 30 that is parallel to $9\mathbf{i} + 12\mathbf{j}$.

B13 Show that $\frac{7}{25}\mathbf{i} + \frac{24}{25}\mathbf{j}$ is a unit vector.

B14 (a) Calculate the magnitude of $12\mathbf{i} + 16\mathbf{j}$.
(b) Hence write down a unit vector in the same direction as $12\mathbf{i} + 16\mathbf{j}$.

Exercise B (answers p 189)

1 Vectors are given by $\mathbf{p} = \begin{bmatrix} 5 \\ 0 \end{bmatrix}$, $\mathbf{q} = \begin{bmatrix} -1 \\ 2 \end{bmatrix}$ and $\mathbf{r} = \begin{bmatrix} -2 \\ -6 \end{bmatrix}$.

Write each of these as a column vector.

(a) $\mathbf{p} + 4\mathbf{q}$ (b) $-\mathbf{p} - 2\mathbf{r}$ (c) $\mathbf{p} + 3\mathbf{q} + \mathbf{r}$

2 Given that $\mathbf{a} = \begin{bmatrix} -2 \\ -3 \end{bmatrix}$ and $\mathbf{b} = \begin{bmatrix} 1 \\ 1 \end{bmatrix}$, calculate each of these.

(a) $|\mathbf{a}|$ (b) $|\mathbf{a} - \mathbf{b}|$ (c) $|-\mathbf{b}|$ (d) $|3\mathbf{b} - 2\mathbf{a}|$

3 Points A, B and C have coordinates $(3, 9)$, $(1, -3)$ and $(-2, -5)$ respectively.

(a) Write $\overrightarrow{CB}$ in terms of $\mathbf{i}$ and $\mathbf{j}$.

(b) Find the distance between points A and B.

4 Find the magnitude of the vector $3(\mathbf{i} - 5\mathbf{j})$.

5 Sort these into three pairs of parallel vectors.

$\mathbf{a} = 5\mathbf{i} - 2\mathbf{j}$ $\mathbf{b} = 2\mathbf{i} + \mathbf{j}$ $\mathbf{c} = 10\mathbf{i} - 6\mathbf{j}$

$\mathbf{d} = \frac{5}{8}\mathbf{i} - \frac{3}{8}\mathbf{j}$ $\mathbf{e} = -10\mathbf{i} + 4\mathbf{j}$ $\mathbf{f} = 10\mathbf{i} + 5\mathbf{j}$

6 If $\mathbf{a} = 4\mathbf{i} + k\mathbf{j}$, where k is a constant, and $|\mathbf{a}| = 2\sqrt{5}$, find the two possible values of k.

7 Write down all the vectors that are parallel to $\begin{bmatrix} 3 \\ 4 \end{bmatrix}$ and have a magnitude of 15.

8 (a) Show that $\frac{5}{13}\mathbf{i} + \frac{12}{13}\mathbf{j}$ is a unit vector.

(b) Find a unit vector that is parallel to $3\mathbf{i} - 4\mathbf{j}$.

9 Vectors $\mathbf{a}$ and $\mathbf{b}$ are given by $\mathbf{a} = k\mathbf{i} + 6\mathbf{j}$ and $\mathbf{b} = 2\mathbf{i} + l\mathbf{j}$, where k and l are constants. Given that $\mathbf{a} + \mathbf{b} = 3\mathbf{i} + 2\mathbf{j}$, find the values of k and l.

10 $PQRS$ is a trapezium where PQ is parallel to SR. The length of SR is twice the length of PQ.

T is the mid-point of QR and the lines PR and ST cross at the point X.

$\overrightarrow{PQ} = \mathbf{a}$ and $\overrightarrow{QR} = \mathbf{b}$.

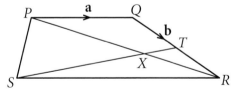

(a) Write each of these in terms of $\mathbf{a}$ and $\mathbf{b}$.

(i) $\overrightarrow{PR}$ (ii) $\overrightarrow{RS}$ (iii) $\overrightarrow{TR}$ (iv) $\overrightarrow{TS}$ (v) $\overrightarrow{PS}$

(b) (i) Show that $\overrightarrow{PX} = k(\mathbf{a} + \mathbf{b})$ for some number k.

(ii) Show that $\overrightarrow{XS} = l(\frac{1}{2}\mathbf{b} - 2\mathbf{a})$ for some number l.

(iii) Use the vector equation $\overrightarrow{PX} + \overrightarrow{XS} = \overrightarrow{PS}$ to find the values of k and l.

C Vectors in three dimensions (answers p 190)

Many situations where we use vectors are three-dimensional.
For example, the velocity of a rocket in space, the displacement of
a person on a rollercoaster and the acceleration of a car can all be
represented by three-dimensional vectors.

The ideas and techniques in sections A and B can be extended to three dimensions.

C1 In the diagram, *ABCD* is a rectangle and *AF*, *BG*, *CH* and *DE* are all parallel.

DE is the same length as *CH*.
AF is the same length as *BG*.
AF is half the length of *DE*.
$\overrightarrow{AD} = \mathbf{p}$, $\overrightarrow{DC} = \mathbf{q}$ and $\overrightarrow{CH} = \mathbf{r}$.

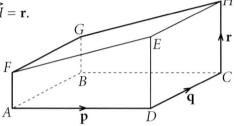

(a) Write these vectors in terms of **p**, **q** and **r**.

   (i) $\overrightarrow{AF}$ (ii) $\overrightarrow{AC}$ (iii) $\overrightarrow{AE}$ (iv) $\overrightarrow{DB}$

   (v) $\overrightarrow{AH}$ (vi) $\overrightarrow{HB}$ (vii) $\overrightarrow{DG}$ (viii) $\overrightarrow{BF}$

(b) Show that the vectors $\overrightarrow{AG}$ and $\overrightarrow{DH}$ are not parallel.

A coordinate system is needed for vectors in three dimensions.
Conventionally we use coordinate axes that are at right angles to each other and
are labelled *x*, *y* and *z* as shown.

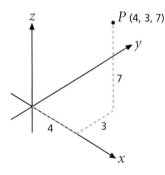

Think of the *x*- and *y*-axes as being drawn
on a horizontal surface (like a table) and
the *z*-axis pointing straight up from it.

The point labelled *P* has coordinates (4, 3, 7).

ABCDEFGH is a cuboid.
The coordinates of *A*, *C* and *G* are shown.

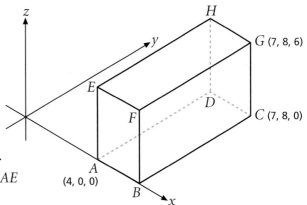

C2 Write down the coordinates of

    (a) *B*   **(b)** *D*

    (c) *F*   **(d)** *H*

C3 Write down the lengths of these edges.

    (a) *AB*   **(b)** *AD*   **(c)** *AE*

This diagram shows the vectors $\overrightarrow{EC}$ and $\overrightarrow{BC}$ in the cuboid.

The vector from *E* to *C* is

- 3 units in the *x*-direction
- 8 units in the *y*-direction
- −6 units in the *z*-direction

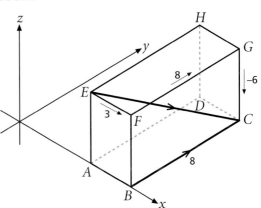

It can be written as a
column vector as $\overrightarrow{EC} = \begin{bmatrix} 3 \\ 8 \\ -6 \end{bmatrix}$ ◄── *x*-component
                        ◄── *y*-component
                        ◄── *z*-component

The vector from *B* to *C* is

- 0 units in the *x*-direction
- 8 units in the *y*-direction
- 0 units in the *z*-direction

and so can be written as $\overrightarrow{BC} = \begin{bmatrix} 0 \\ 8 \\ 0 \end{bmatrix}$.

C4 Write each of these as a column vector.

    (a) $\overrightarrow{AF}$ **(b)** $\overrightarrow{DG}$ **(c)** $\overrightarrow{AC}$ **(d)** $\overrightarrow{BE}$

    (e) $\overrightarrow{EA}$ **(f)** $\overrightarrow{AG}$ **(g)** $\overrightarrow{BH}$ **(h)** $\overrightarrow{FD}$

C5 Sort these into two groups of parallel vectors.

$$\begin{bmatrix} 2 \\ -1 \\ 3 \end{bmatrix}, \begin{bmatrix} -6 \\ 3 \\ -9 \end{bmatrix}, \begin{bmatrix} 12 \\ -6 \\ 16 \end{bmatrix}, \begin{bmatrix} 4 \\ -2 \\ 6 \end{bmatrix}, \begin{bmatrix} 6 \\ -3 \\ 8 \end{bmatrix}$$

In three dimensions, unit vectors **i**, **j** and **k** are defined as

$$\mathbf{i} = \begin{bmatrix} 1 \\ 0 \\ 0 \end{bmatrix}, \ \mathbf{j} = \begin{bmatrix} 0 \\ 1 \\ 0 \end{bmatrix}, \ \mathbf{k} = \begin{bmatrix} 0 \\ 0 \\ 1 \end{bmatrix}$$

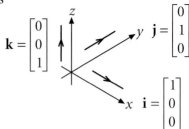

Hence any vector can be written as a linear combination of **i**, **j** and **k**.

For example, $\begin{bmatrix} 4 \\ -1 \\ 7 \end{bmatrix} = 4\begin{bmatrix} 1 \\ 0 \\ 0 \end{bmatrix} - \begin{bmatrix} 0 \\ 1 \\ 0 \end{bmatrix} + 7\begin{bmatrix} 0 \\ 0 \\ 1 \end{bmatrix} = 4\mathbf{i} - \mathbf{j} + 7\mathbf{k}.$

C6 Write $\begin{bmatrix} 2 \\ 6 \\ -5 \end{bmatrix}$ in terms of **i**, **j** and **k**.

C7 Write $3\mathbf{i} - 2\mathbf{j} + 8\mathbf{k}$ as a column vector.

C8 Show that the vectors $\mathbf{i} + 2\mathbf{j} - 3\mathbf{k}$ and $2\mathbf{i} + 4\mathbf{j} - 6\mathbf{k}$ are parallel.

C9 This diagram shows the vector $\mathbf{i} + 3\mathbf{j} + 2\mathbf{k}$ with the components shown by the dotted lines. A triangle is shaded.

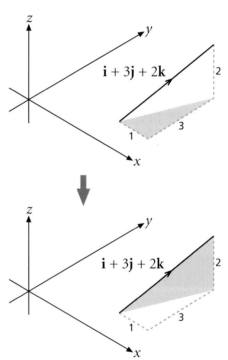

(a) Show that the shaded triangle is a right-angled triangle.

(b) What is the length (in surd form) of the hypotenuse of the shaded triangle?

In the next diagram, a second triangle is shaded.

(c) Is the shaded triangle right-angled?

(d) Use this triangle to work out the magnitude of the vector $\mathbf{i} + 3\mathbf{j} + 2\mathbf{k}$, giving your result in surd form.

C10 Work out the magnitude of the vector $\begin{bmatrix} 5 \\ 12 \\ 3 \end{bmatrix}$.

C11 This diagram shows the vector $\begin{bmatrix} 4 \\ -1 \\ -5 \end{bmatrix}$.

The components are shown by the dotted lines.

Work out the magnitude of the vector.

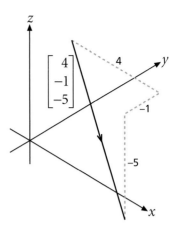

This diagram shows the vector $a\mathbf{i} + b\mathbf{j} + c\mathbf{k}$ with the components shown by the dotted lines.

The shaded triangle is right-angled so, by Pythagoras's theorem, the length of its hypotenuse is $\sqrt{a^2 + b^2}$.

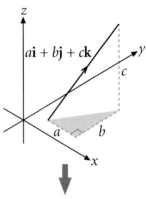

The second shaded triangle is right-angled too so, using Pythagoras's theorem again, the magnitude of the vector $a\mathbf{i} + b\mathbf{j} + c\mathbf{k}$ is

$$\sqrt{\left(\sqrt{a^2 + b^2}\right)^2 + c^2} = \sqrt{a^2 + b^2 + c^2}.$$

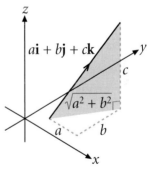

This process will give the magnitude of any vector, including a vector with one or more negative components.

For example, the magnitude of the vector $\begin{bmatrix} 4 \\ -1 \\ -5 \end{bmatrix}$ is $\sqrt{4^2 + (-1)^2 + (-5)^2} = \sqrt{16 + 1 + 25}$

$$= \sqrt{42}$$

The magnitude of any vector $a\mathbf{i} + b\mathbf{j} + c\mathbf{k}$ or $\begin{bmatrix} a \\ b \\ c \end{bmatrix}$ is $\sqrt{a^2 + b^2 + c^2}$.

C12 Find the magnitude of each of these vectors.

(a) $\mathbf{i} - 2\mathbf{j} + 2\mathbf{k}$ (b) $3\mathbf{i} + \mathbf{j} - 5\mathbf{k}$ (c) $\begin{bmatrix} 0 \\ -3 \\ -6 \end{bmatrix}$

C13 *ABCDE* is a square-based pyramid.
The coordinates of points *A* to *E* are shown.

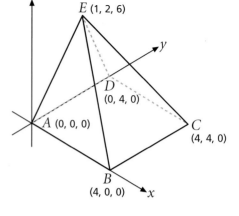

(a) Write each of these as a column vector.

(i) $\overrightarrow{AB}$ (ii) $\overrightarrow{AE}$ (iii) $\overrightarrow{EA}$

(b) By considering the vector addition
$$\overrightarrow{EB} = \overrightarrow{EA} + \overrightarrow{AB}$$
write $\overrightarrow{EB}$ as a column vector.

(c) Write each of these in terms of **i**, **j** and **k**.

(i) $\overrightarrow{AC}$ (ii) $\overrightarrow{EC}$ (iii) $\overrightarrow{DE}$

(d) Find $|\overrightarrow{DE}|$.

C14 Vectors are given by $\mathbf{a} = \begin{bmatrix} -2 \\ 3 \\ 0 \end{bmatrix}$, $\mathbf{b} = \begin{bmatrix} 4 \\ -1 \\ 7 \end{bmatrix}$ and $\mathbf{c} = \begin{bmatrix} 2 \\ 0 \\ -4 \end{bmatrix}$.

(a) Write each of these as a column vector.

(i) $\mathbf{a} + \mathbf{b}$ (ii) $3\mathbf{c} + 2\mathbf{b}$ (iii) $4\mathbf{b} - \mathbf{c}$ (iv) $\frac{1}{2}\mathbf{c} - 2\mathbf{a}$ (v) $2(\mathbf{a} + 2\mathbf{c})$

(b) Evaluate $|\mathbf{a} + \mathbf{b} + \mathbf{c}|$.

C15 Vectors **p**, **q** and **r** are given by $\mathbf{p} = \mathbf{i} + \mathbf{j} + 4\mathbf{k}$, $\mathbf{q} = -2\mathbf{i} + 5\mathbf{j} + \mathbf{k}$ and $\mathbf{r} = 4\mathbf{i} - \mathbf{k}$.

(a) Show that $|\mathbf{p}| = 3\sqrt{2}$.

(b) Write the following in terms of **i**, **j** and **k**.

(i) $\mathbf{q} + \mathbf{r}$ (ii) $\mathbf{p} - \mathbf{q}$ (iii) $2\mathbf{q} + \mathbf{r}$

(c) Evaluate $|\mathbf{q} + \mathbf{r}|$ in surd form.

Exercise C (answers p 190)

1 Vectors are given by $\mathbf{a} = \begin{bmatrix} 3 \\ 0 \\ -4 \end{bmatrix}$, $\mathbf{b} = \begin{bmatrix} -1 \\ -5 \\ 2 \end{bmatrix}$ and $\mathbf{c} = \begin{bmatrix} -1 \\ 10 \\ 0 \end{bmatrix}$.

(a) Write each of these as a column vector.

(i) $\mathbf{c} - \mathbf{b}$ (ii) $2(\mathbf{a} + 3\mathbf{b})$ (iii) $\mathbf{a} + 2\mathbf{b} + \mathbf{c}$

(b) Evaluate $|2\mathbf{b} + \mathbf{c}|$.

2 (a) Find the magnitude of the vector $\mathbf{i} + 4\mathbf{j} - 8\mathbf{k}$.

(b) Hence write down the magnitude of the vector $3\mathbf{i} + 12\mathbf{j} - 24\mathbf{k}$.

3 (a) Find the magnitude of the vector $2\mathbf{i} + 2\mathbf{j} - \mathbf{k}$.

(b) Hence write down a unit vector parallel to $2\mathbf{i} + 2\mathbf{j} - \mathbf{k}$.

4 *ABCDEFGH* is a prism.

The coordinates of points *A* to *D* are shown on the diagram.

Vectors $\overrightarrow{AE}$, $\overrightarrow{BF}$, $\overrightarrow{CG}$ and $\overrightarrow{DH}$ are parallel to the *z*-axis and have magnitude 2.

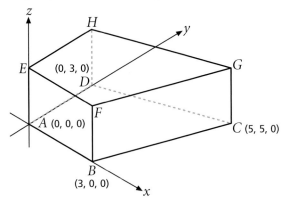

(a) Write each of these as a column vector.

   (i) $\overrightarrow{AB}$ (ii) $\overrightarrow{AE}$ (iii) $\overrightarrow{EH}$

(b) By considering the vector addition $\overrightarrow{HB} = \overrightarrow{HE} + \overrightarrow{EA} + \overrightarrow{AB}$, write $\overrightarrow{HB}$ as a column vector.

(c) Write each of these in terms of **i**, **j** and **k**.

   (i) $\overrightarrow{BG}$ (ii) $\overrightarrow{HC}$ (iii) $\overrightarrow{AG}$ (iv) $\overrightarrow{CE}$

(d) Work out the magnitude of $\overrightarrow{DF}$.

5 Sort these into three pairs of parallel vectors.

$$\mathbf{u} = \begin{bmatrix} 1 \\ 3 \\ -4 \end{bmatrix}, \quad \mathbf{v} = \begin{bmatrix} 4 \\ 12 \\ -4 \end{bmatrix}, \quad \mathbf{w} = \begin{bmatrix} 3 \\ 9 \\ 12 \end{bmatrix}, \quad \mathbf{x} = \begin{bmatrix} \frac{1}{2} \\ 1\frac{1}{2} \\ 2 \end{bmatrix}, \quad \mathbf{y} = \begin{bmatrix} 2 \\ 6 \\ -8 \end{bmatrix}, \quad \mathbf{z} = \begin{bmatrix} -1 \\ -3 \\ 1 \end{bmatrix}$$

6 Vectors **a** and **b** are given by $\mathbf{a} = 2\mathbf{i} - \mathbf{j} + 3\mathbf{k}$ and $\mathbf{b} = \mathbf{i} - 4\mathbf{j} + 4\mathbf{k}$.

Find the vector $4\mathbf{a} - 3\mathbf{b}$ in terms of **i**, **j** and **k**.

7 Vectors **p** and **q** are parallel.

They are given by $\mathbf{p} = x\mathbf{i} - 9\mathbf{j} + 6\mathbf{k}$ and $\mathbf{q} = 2\mathbf{i} + y\mathbf{j} + 4\mathbf{k}$, where *x* and *y* are constants. Find the values of *x* and *y*.

8 The vector $5\mathbf{i} + \alpha\mathbf{j} + 2\mathbf{k}$ has a magnitude of $3\sqrt{5}$.

Find the two possible values of α.

9 Write down a vector that is parallel to $2\mathbf{i} - 3\mathbf{j} - 6\mathbf{k}$ and has a magnitude of 14.

*10 Vectors **a**, **b** and **c** are given by $\mathbf{a} = \mathbf{i} + \mathbf{j} - 2\mathbf{k}$, $\mathbf{b} = 5\mathbf{i} - \mathbf{j} + 3\mathbf{k}$ and $\mathbf{c} = -2\mathbf{i} + \mathbf{j} + \mathbf{k}$.

Show that it is not possible to write **a** as a linear combination of **b** and **c**.

D Position vectors in two and three dimensions (answers p 191)

For every point P there is associated a unique vector $\overrightarrow{OP}$, where O is the origin.
The vector $\overrightarrow{OP}$ is called the **position vector** of point P and is sometimes written as **p**.

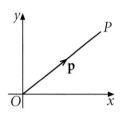

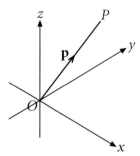

In two dimensions, the point (x, y) has position vector $\begin{bmatrix} x \\ y \end{bmatrix}$ or $x\mathbf{i} + y\mathbf{j}$.

In three dimensions, the point (x, y, z) has position vector $\begin{bmatrix} x \\ y \\ z \end{bmatrix}$ or $x\mathbf{i} + y\mathbf{j} + z\mathbf{k}$.

For example, the point with coordinates $(2, 3)$ has
position vector $\begin{bmatrix} 2 \\ 3 \end{bmatrix}$ or $2\mathbf{i} + 3\mathbf{j}$.

The position vector is fixed with its 'tail' at the origin as shown.

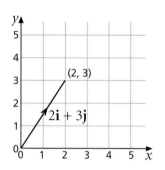

A **free vector** is not fixed to an origin and can be represented by any line of the correct magnitude and direction.

For example, each of these represents the free vector $2\mathbf{i} + 3\mathbf{j}$.

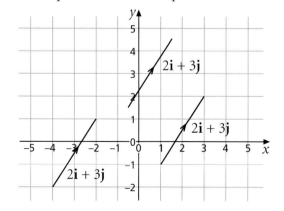

A vector can be used as a position vector or a free vector.

In this diagram point A has position vector **a** and point B has position vector **b**.
$OAXB$ is a parallelogram.

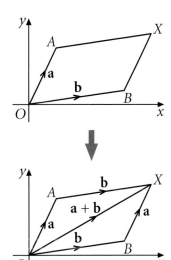

$OAXB$ is a parallelogram so $\overrightarrow{BX} = $ **a** and $\overrightarrow{AX} = $ **b**.

The position vector of X is $\overrightarrow{OX}$ which is equivalent to $\overrightarrow{OA} + \overrightarrow{AX}$ so the position vector of X can be written as **a** + **b**.

D1 Points P and Q have position vectors **p** and **q** as shown.

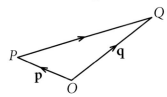

Use the vector addition $\overrightarrow{PQ} = \overrightarrow{PO} + \overrightarrow{OQ}$ to write the vector $\overrightarrow{PQ}$ in terms of **p** and **q**.

The relationship found in D1 is often useful and can be stated as follows.

K If point A has position vector **a** (or $\overrightarrow{OA}$) and point B has position vector **b** (or $\overrightarrow{OB}$), then
$\overrightarrow{AB} = $ **b** − **a** (or $\overrightarrow{OB} - \overrightarrow{OA}$).

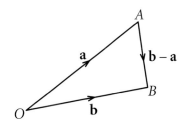

D2 Two points X and Y have position vectors $\mathbf{x} = \mathbf{i} + 6\mathbf{j}$ and $\mathbf{y} = 2\mathbf{i} - 7\mathbf{j}$ respectively.
  (a) Draw a sketch to show the position of points X and Y.
  (b) What are the coordinates of the point with position vector $\mathbf{x} + \mathbf{y}$?
  (c) Use the relationship $\overrightarrow{XY} = \mathbf{y} - \mathbf{x}$ to write the vector $\overrightarrow{XY}$ in terms of **i** and **j**.

D3 Points M and N have position vectors $\mathbf{m}$ and $\mathbf{n}$ as shown.
P is the point on MN such that $MP = \frac{1}{4}MN$.

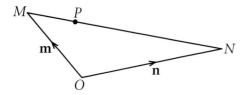

(a) Write $\overrightarrow{MP}$ in terms of $\mathbf{m}$ and $\mathbf{n}$.

(b) Show that the position vector of the point P in terms of $\mathbf{m}$ and $\mathbf{n}$ is $\frac{3}{4}\mathbf{m} + \frac{1}{4}\mathbf{n}$.

D4 Two points P and Q have position vectors $\mathbf{p} = \begin{bmatrix} 2 \\ -3 \\ 4 \end{bmatrix}$ and $\mathbf{q} = \begin{bmatrix} 0 \\ 5 \\ -2 \end{bmatrix}$.

(a) Write $\overrightarrow{PQ}$ as a column vector.

(b) Calculate the distance between points P and Q.

In general, if two points A and B have position vectors $\mathbf{a} = \begin{bmatrix} x_1 \\ y_1 \\ z_1 \end{bmatrix}$ and $\mathbf{b} = \begin{bmatrix} x_2 \\ y_2 \\ z_2 \end{bmatrix}$

then $\overrightarrow{AB} = \mathbf{b} - \mathbf{a} = \begin{bmatrix} x_2 - x_1 \\ y_2 - y_1 \\ z_2 - z_1 \end{bmatrix}$

and $|\overrightarrow{AB}| = \sqrt{(x_2 - x_1)^2 + (y_2 - y_1)^2 + (z_2 - z_1)^2}$

which is the distance between points A and B.

Example 2

The points P and Q have position vectors $\mathbf{p} = 2\mathbf{i} - \mathbf{j} - 8\mathbf{k}$ and $\mathbf{q} = 5\mathbf{i} - \mathbf{j} + 4\mathbf{k}$ respectively.
The point X divides PQ such that $PX = \frac{1}{3}PQ$.
Find $|\overrightarrow{OX}|$.

Solution

$\overrightarrow{PQ} = \mathbf{q} - \mathbf{p} = (5\mathbf{i} - \mathbf{j} + 4\mathbf{k}) - (2\mathbf{i} - \mathbf{j} - 8\mathbf{k})$

$\qquad = 3\mathbf{i} + 12\mathbf{k}$

$\overrightarrow{OX} = \overrightarrow{OP} + \overrightarrow{PX} = \overrightarrow{OP} + \frac{1}{3}\overrightarrow{PQ}$

$\qquad = (2\mathbf{i} - \mathbf{j} - 8\mathbf{k}) + \frac{1}{3}(3\mathbf{i} + 12\mathbf{k})$

$\qquad = 2\mathbf{i} - \mathbf{j} - 8\mathbf{k} + \mathbf{i} + 4\mathbf{k}$

$\qquad = 3\mathbf{i} - \mathbf{j} - 4\mathbf{k}$

So $|\overrightarrow{OX}| = \sqrt{3^2 + (-1)^2 + (-4)^2} = \sqrt{26}$

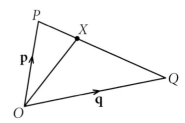

It is helpful to draw a diagram but it doesn't have to look three-dimensional.

Exercise D (answers p 191)

1 The points A and B have position vectors **a** and **b** respectively.
The point P divides AB such that $AP = \frac{1}{5}AB$.

Find, in terms of **a** and **b**, the position vector of P.

2 Two points P and Q have position vectors $\mathbf{p} = 2\mathbf{i} - \mathbf{j} - 8\mathbf{k}$ and $\mathbf{q} = 3\mathbf{i} + 4\mathbf{j} - \mathbf{k}$.

(a) Write $\overrightarrow{PQ}$ as a vector in terms of **i**, **j** and **k**.

(b) Calculate the distance between P and Q.

3 $OABCDEFG$ is a cube with edges of length 4.
A, C and D are points on the x-, y- and z-axes respectively.

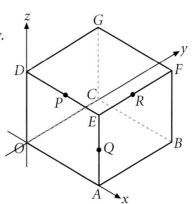

(a) Using column vector notation, write down
the position vector of each vertex of the cube.

(b) Find the vectors $\overrightarrow{AG}$ and $\overrightarrow{BD}$.

P, Q and R are the mid-points of edges DE, AE and EF.

(c) Find the position vectors of P, Q and R.

(d) (i) Find the vectors $\overrightarrow{PQ}$, $\overrightarrow{QR}$ and $\overrightarrow{RP}$.

(ii) Calculate the vector addition $\overrightarrow{PQ} + \overrightarrow{QR} + \overrightarrow{RP}$ and comment on your result.

4 Points A, B and C have position vectors

$$\mathbf{a} = 5\mathbf{i} + 5\mathbf{j} + 4\mathbf{k} \qquad \mathbf{b} = 4\mathbf{i} - 7\mathbf{j} + \mathbf{k} \qquad \mathbf{c} = 7\mathbf{i} + 4\mathbf{j} - \mathbf{k}$$

Show that points A, B and C are all on the surface of a sphere whose centre is $(0, 0, 0)$.

5 Points P, Q, R and S have position vectors

$$\mathbf{p} = 3\mathbf{i} - \mathbf{j} + 2\mathbf{k} \qquad \mathbf{q} = 2\mathbf{i} + \mathbf{j} + 4\mathbf{k} \qquad \mathbf{r} = 5\mathbf{i} + 2\mathbf{j} - 3\mathbf{k} \qquad \mathbf{s} = 7\mathbf{i} - 2\mathbf{j} - 7\mathbf{k}$$

Find the vectors $\overrightarrow{PQ}$ and $\overrightarrow{RS}$ and show that they are parallel.

6 Find the distance between each of these pairs of points.

(a) $A(1, -6, 3)$ and $B(-8, 0, 7)$ (b) $V(5, -4, -7)$ and $W(12, -3, 1)$

7 The points A, B and C have position vectors $\mathbf{i} - 2\mathbf{j} + \mathbf{k}$, $10\mathbf{i} + \mathbf{j} + 4\mathbf{k}$ and $4\mathbf{i} - \mathbf{j} + 2\mathbf{k}$ respectively. Prove that A, B and C lie on the same straight line.

8 The coordinates of P and Q are $(4, -8, 1)$ and $(2, -2, k)$ respectively.
Given that $|\overrightarrow{PQ}| = 7$, find the possible values of k.

E The vector equation of a line (answers p 192)

A set of points is defined so that the position vector of each point in the set has the form

$$\begin{bmatrix} x \\ y \end{bmatrix} = \begin{bmatrix} 1 \\ 6 \end{bmatrix} + t \begin{bmatrix} 2 \\ -1 \end{bmatrix} \text{ for all real values of } t$$

For example, when $t = 2$, the position vector is given by $\begin{bmatrix} x \\ y \end{bmatrix} = \begin{bmatrix} 1 \\ 6 \end{bmatrix} + 2 \begin{bmatrix} 2 \\ -1 \end{bmatrix} = \begin{bmatrix} 5 \\ 4 \end{bmatrix}$.

Hence the point with coordinates (5, 4) is in the set.

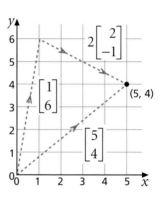

E1 (a) Show that $t = 5$ gives the point with position vector $\begin{bmatrix} 11 \\ 1 \end{bmatrix}$.

(b) Find the position vector given by each of these values of t.

    (i) $t = 0$ (ii) $t = 1$ (iii) $t = -1$ (iv) $t = 3$

(c) (i) On the same diagram, plot all points defined by the position vectors found so far.

    (ii) Explain why all the points lie on a straight line.

    (iii) Find the equation of this straight line in the form $ax + by = c$.

All the points given by the expression $\begin{bmatrix} 1 \\ 6 \end{bmatrix} + t \begin{bmatrix} 2 \\ -1 \end{bmatrix}$ lie on a straight line.

Let **r** be the position vector of any point on the line.

Then the equation $\mathbf{r} = \begin{bmatrix} 1 \\ 6 \end{bmatrix} + t \begin{bmatrix} 2 \\ -1 \end{bmatrix}$ is called a **vector equation** of the line.

As t is a number that varies, it is sometimes called a **scalar parameter**.

We can also use **i, j, k** notation to write down the vector equation of a line.

For example, we could write the vector equation above in the form

    $\mathbf{r} = \mathbf{i} + 6\mathbf{j} + t(2\mathbf{i} - \mathbf{j})$

E2 A line has the vector equation $\mathbf{r} = \mathbf{i} - \mathbf{j} + t(\mathbf{i} + 2\mathbf{j})$, where t is a scalar parameter.

(a) Show that $t = 3$ gives the point with position vector $4\mathbf{i} + 5\mathbf{j}$.

(b) Find the position vector given by each of these values of t.

    (i) $t = 1$ (ii) $t = 0$ (iii) $t = -3$ (iv) $t = 5$

(c) (i) Draw the straight line through the points defined by the vector equation.

    (ii) Find the equation of this straight line.

E3 A line has the vector equation $\mathbf{r} = 2\mathbf{i} + \mathbf{j} + s(-\mathbf{i} - 2\mathbf{j})$, where s is a scalar parameter.

(a) Find the position vector given by each of these values of s.

    (i) $s = 1$ (ii) $s = 0$ (iii) $s = -3$

(b) (i) Draw the straight line through the points defined by the vector equation.

    (ii) Compare this with the line drawn in E2(c).

    (iii) Can you explain what you find?

K   The vector equation for a straight line is not unique.

For example, both the following diagrams show the straight line with equation $y = x + 3$.

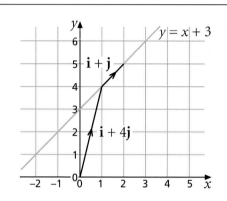

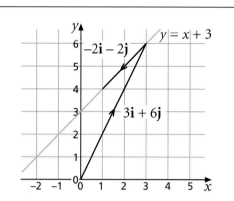

Each point on the line has a position vector that can be written in the form

$$\mathbf{i} + 4\mathbf{j} + s(\mathbf{i} + \mathbf{j})$$

So $\mathbf{r} = \mathbf{i} + 4\mathbf{j} + s(\mathbf{i} + \mathbf{j})$ is a vector equation of the line.

Each point on the line has a position vector that can be written in the form

$$3\mathbf{i} + 6\mathbf{j} + t(-2\mathbf{i} - 2\mathbf{j})$$

So $\mathbf{r} = 3\mathbf{i} + 6\mathbf{j} + t(-2\mathbf{i} - 2\mathbf{j})$ is a vector equation of the line.

E4 Give a vector equation for each of these lines.

(a) $y = x + 4$ (b) $y = 6 - x$ (c) $y = 2x + 1$

E5 (a) On the same diagram, sketch the lines l_1 and l_2 with the following vector equations.

$$l_1: \mathbf{r} = \begin{bmatrix} 1 \\ 3 \end{bmatrix} + \lambda \begin{bmatrix} 1 \\ 2 \end{bmatrix} \qquad l_2: \mathbf{r} = \begin{bmatrix} 5 \\ 2 \end{bmatrix} + \mu \begin{bmatrix} -2 \\ -4 \end{bmatrix}$$

> The Greek letters λ (lambda) and μ (mu) are often used for scalar parameters in vector work.

(b) (i) What do you notice about your lines?

    (ii) How could you have predicted this from the vector equations?

E6 Which of the lines whose vector equations are given below is parallel to the line with vector equation $\mathbf{r} = \mathbf{j} + \lambda(2\mathbf{i} + 3\mathbf{j})$?

$L_1: \mathbf{r} = 5\mathbf{j} + s(3\mathbf{i} + 4\mathbf{j})$ $L_2: \mathbf{r} = 2\mathbf{i} + 7\mathbf{j} + t(4\mathbf{i} + 6\mathbf{j})$ $L_3: \mathbf{r} = -2\mathbf{i} - 3\mathbf{j} + \mu(\mathbf{i} + \mathbf{j})$

The vector equation of a line in three dimensions is given in a similar way.

The diagram below shows the line with vector equation

$$\mathbf{r} = \begin{bmatrix} 4 \\ 3 \\ 7 \end{bmatrix} + \lambda \begin{bmatrix} 1 \\ 0 \\ -3 \end{bmatrix}$$

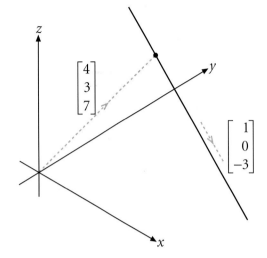

It is more difficult to visualise a line in three dimensions.

E7 **(a)** Write down the coordinates of three points on the line above.

(b) The point $(9, 3, -8)$ lies on the line.
What value of λ corresponds to this point?

(c) Show that the point with position vector $\begin{bmatrix} 7 \\ 3 \\ -1 \end{bmatrix}$ does not lie on the line.

E8 The vector equation of a line is given by
$$\mathbf{r} = \mathbf{i} + 2\mathbf{j} - 3\mathbf{k} + \lambda(2\mathbf{i} - 4\mathbf{j} + \mathbf{k})$$

(a) What position vector is given by $\lambda = 3$?

(b) What value of λ gives the position vector $-\mathbf{i} + 6\mathbf{j} - 4\mathbf{k}$?

(c) Show that the point with position vector $3\mathbf{i} - 3\mathbf{j} - 2\mathbf{k}$ does not lie on the line.

K If a point P with position vector $\mathbf{p}$ is on a line and $\mathbf{d}$ is a vector parallel to the line (called a **direction vector**) then a vector equation of the line is $\mathbf{r} = \mathbf{p} + t\mathbf{d}$, where t is a scalar parameter.

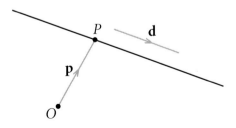

The general form of a vector equation of a line is

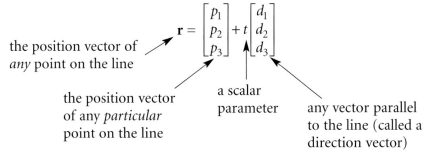

the position vector of *any* point on the line

$$\mathbf{r} = \begin{bmatrix} p_1 \\ p_2 \\ p_3 \end{bmatrix} + t \begin{bmatrix} d_1 \\ d_2 \\ d_3 \end{bmatrix}$$

the position vector of any *particular* point on the line

a scalar parameter

any vector parallel to the line (called a direction vector)

Using $\mathbf{i}, \mathbf{j}, \mathbf{k}$ notation this equation is

$$\mathbf{r} = (p_1\mathbf{i} + p_2\mathbf{j} + p_3\mathbf{k}) + t(d_1\mathbf{i} + d_2\mathbf{j} + d_3\mathbf{k})$$

Lines are parallel if their direction vectors are parallel.

If two points A and B with position vectors $\mathbf{a}$ and $\mathbf{b}$ are on a line then $\mathbf{b} - \mathbf{a}$ is a direction vector for that line and a vector equation is $\mathbf{r} = \mathbf{a} + t(\mathbf{b} - \mathbf{a})$.

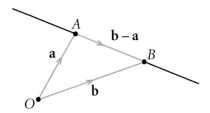

Example 3

Points A and B have position vectors $\mathbf{i} + 2\mathbf{j} - 5\mathbf{k}$ and $-5\mathbf{i} + 6\mathbf{j} - 3\mathbf{k}$ respectively.

Find, in vector form, the equation of the straight line that passes through A and B.

Solution

The vector $\overrightarrow{AB}$ is a direction vector for the line.

$\overrightarrow{AB} = (-5\mathbf{i} + 6\mathbf{j} - 3\mathbf{k}) - (\mathbf{i} + 2\mathbf{j} - 5\mathbf{k})$ *This is the position vector of B – the position vector of A.*

$\quad\quad = -6\mathbf{i} + 4\mathbf{j} + 2\mathbf{k}$ *You could use column vectors in your working.*

A is a point on the line with position vector $\mathbf{i} + 2\mathbf{j} - 5\mathbf{k}$, so a vector equation is

$\mathbf{r} = \mathbf{i} + 2\mathbf{j} - 5\mathbf{k} + t(-6\mathbf{i} + 4\mathbf{j} + 2\mathbf{k})$

You could have written this using column vectors as $\mathbf{r} = \begin{bmatrix} 1 \\ 2 \\ -5 \end{bmatrix} + t \begin{bmatrix} -6 \\ 4 \\ 2 \end{bmatrix}$.

Any scalar multiple of $\overrightarrow{AB}$ is a direction vector for the line so, for example, instead of the direction vector $-6\mathbf{i} + 4\mathbf{j} + 2\mathbf{k}$ you could have used $-3\mathbf{i} + 2\mathbf{j} + \mathbf{k}$ to obtain
$\mathbf{r} = \mathbf{i} + 2\mathbf{j} - 5\mathbf{k} + s(-3\mathbf{i} + 2\mathbf{j} + \mathbf{k})$.

Sometimes it can be useful to write a vector equation in a slightly different form, as the examples on the next page show.

Example 4

A line has a vector equation given by $\mathbf{r} = -2\mathbf{i} + \mathbf{j} + 5\mathbf{k} + \lambda(\mathbf{i} - 3\mathbf{j} + 2\mathbf{k})$.

Show that the point P with position vector $3\mathbf{i} - 14\mathbf{j} + 15\mathbf{k}$ is on this line.

Solution

Rewrite the vector equation.

$\mathbf{r} = -2\mathbf{i} + \mathbf{j} + 5\mathbf{k} + \lambda(\mathbf{i} - 3\mathbf{j} + 2\mathbf{k})$

$\quad = -2\mathbf{i} + \mathbf{j} + 5\mathbf{k} + \lambda\mathbf{i} - 3\lambda\mathbf{j} + 2\lambda\mathbf{k}$

$\quad = (-2 + \lambda)\mathbf{i} + (1 - 3\lambda)\mathbf{j} + (5 + 2\lambda)\mathbf{k}$

We need to find a value of λ such that

$\quad (-2 + \lambda)\mathbf{i} + (1 - 3\lambda)\mathbf{j} + (5 + 2\lambda)\mathbf{k} = 3\mathbf{i} - 14\mathbf{j} + 15\mathbf{k}$

Equating coefficients of $\mathbf{i}$ gives

$\quad -2 + \lambda = 3$

$\Rightarrow \qquad \lambda = 5$

This gives $\ 1 - 3\lambda = 1 - 3\times 5 = -14$ which is the required coefficient of $\mathbf{j}$

$\qquad$ and $\ 5 + 2\lambda = 5 + 2\times 5 = 15$ which is the required coefficient of $\mathbf{k}$

Hence P is on the line.

A vector equation of a line can be given in the form $\begin{bmatrix} x \\ y \\ z \end{bmatrix} = \begin{bmatrix} p_1 \\ p_2 \\ p_3 \end{bmatrix} + \lambda \begin{bmatrix} d_1 \\ d_2 \\ d_3 \end{bmatrix}$.

Example 5

A line has the vector equation $\begin{bmatrix} x \\ y \\ z \end{bmatrix} = \begin{bmatrix} -1 \\ 1 \\ -3 \end{bmatrix} + t\begin{bmatrix} 2 \\ -2 \\ 5 \end{bmatrix}$, where t is a scalar parameter.

The point P lies on this line and has coordinates $(5, a, b)$, where a and b are constants. Find the values of a and b.

Solution

$\begin{bmatrix} x \\ y \\ z \end{bmatrix} = \begin{bmatrix} -1 \\ 1 \\ -3 \end{bmatrix} + t\begin{bmatrix} 2 \\ -2 \\ 5 \end{bmatrix} = \begin{bmatrix} -1 + 2t \\ 1 - 2t \\ -3 + 5t \end{bmatrix}$

P has position vector $\begin{bmatrix} 5 \\ a \\ b \end{bmatrix}$ and lies on the line so $\begin{bmatrix} -1 + 2t \\ 1 - 2t \\ -3 + 5t \end{bmatrix} = \begin{bmatrix} 5 \\ a \\ b \end{bmatrix}$ for some t.

Equating the x-values, $-1 + 2t = 5$ which gives $t = 3$.

When $t = 3$, $a = 1 - 2t = -5$

$\qquad\qquad\qquad b = -3 + 5t = 12$.

Hence $a = -5$ and $b = 12$.

Exercise E (answers p 193)

1 In two dimensions, find vector equations for

   (a) the line joining the points $(0, 5)$ and $(1, 9)$

   (b) the x-axis

   (c) the line joining the points with position vectors $5\mathbf{i} - 2\mathbf{j}$ and $6\mathbf{i} - 7\mathbf{j}$

2 In three dimensions, find vector equations for

   (a) the line through the point $(-1, 3, 5)$, parallel to the vector $\begin{bmatrix} -1 \\ 1 \\ -3 \end{bmatrix}$

   (b) the line through the points $(2, 1, 0)$ and $(3, 4, 4)$

   (c) the z-axis

   (d) the line joining the points with position vectors $2\mathbf{i} - 4\mathbf{j} + 7\mathbf{k}$ and $-\mathbf{i} + 2\mathbf{j} - \mathbf{k}$

3 A line has a vector equation given by $\mathbf{r} = \mathbf{i} - 3\mathbf{j} + 2\mathbf{k} + \lambda(2\mathbf{i} + 4\mathbf{j} - 6\mathbf{k})$.
   Show that the point with position vector $2\mathbf{i} - \mathbf{j} - \mathbf{k}$ is on this line.

4 Points A and B have position vectors $3\mathbf{i} + \mathbf{j} + 2\mathbf{k}$ and $11\mathbf{i} - \mathbf{j} - 2\mathbf{k}$.

   (a) Find, in vector form, an equation of the line that passes through A and B.

   (b) Show that the line intersects the x-axis.

5 Points P and Q have position vectors $-2\mathbf{i} + \mathbf{j}$ and $4\mathbf{i} + 4\mathbf{k}$.

   (a) Find, in vector form, an equation of the line that passes through P and Q.

   (b) Show that the line does not intersect the y-axis.

6 Referred to a fixed origin O (an airport), the position vector of aircraft A,
   t minutes after take-off, is given by the vector equation

$$\begin{bmatrix} x \\ y \\ z \end{bmatrix} = t \begin{bmatrix} 1 \\ 5 \\ 0.8 \end{bmatrix}$$

   where the position of the aircraft is x km east and y km north, at a height of z km.

   (a) Find the position vector of aircraft A after 10 minutes.

   The position vector of aircraft B at the same time, t minutes, is given by
   the vector equation

$$\begin{bmatrix} x \\ y \\ z \end{bmatrix} = \begin{bmatrix} 5 \\ 0 \\ 4 \end{bmatrix} + t \begin{bmatrix} 0 \\ 5 \\ 0 \end{bmatrix}$$

   (b) After 3 minutes, how far is aircraft B from O, correct to the nearest 0.1 km?

   (c) How far apart are the aircraft after 4 minutes?

   (d) How far apart are the aircraft after 5 minutes?
   What does this mean?

7 The points A and B have coordinates $(3, 0, -1)$ and $(-5, 2, 1)$ respectively. The point C lies on the line through these points and has coordinates $(p, 5, q)$, where p and q are constants. Find the values of p and q.

8 A vector equation for a straight line is $\begin{bmatrix} x \\ y \end{bmatrix} = \begin{bmatrix} 4 \\ 7 \end{bmatrix} + \lambda \begin{bmatrix} 1 \\ -2 \end{bmatrix}$.

The line intersects a circle that has centre $(0, 0)$ and radius $5\sqrt{2}$. Find the coordinates of the points of intersection.

9 A vector equation for a straight line is $\mathbf{r} = 4\mathbf{i} + \mathbf{j} + 2\mathbf{k} + \lambda(\mathbf{i} + \mathbf{j})$. The line intersects a hollow sphere that has centre $(0, 0, 0)$ and radius 3. Find the coordinates of the points of intersection.

F Intersecting lines (answers p 193)

In two dimensions, there are three possible relationships between two straight lines.

They are parallel.   They intersect at one point.   They are the same line.

D

F1 Two lines are given by these vector equations, where λ and μ are scalar parameters.

$$\begin{bmatrix} x \\ y \end{bmatrix} = \begin{bmatrix} 1 \\ -3 \end{bmatrix} + \lambda \begin{bmatrix} 1 \\ 1 \end{bmatrix} \qquad \begin{bmatrix} x \\ y \end{bmatrix} = \begin{bmatrix} 11 \\ 1 \end{bmatrix} + \mu \begin{bmatrix} 1 \\ -2 \end{bmatrix}$$

(a) Show that, if the lines intersect, there must be values of λ and μ that satisfy these simultaneous equations.

$$1 + \lambda = 11 + \mu$$
$$-3 + \lambda = 1 - 2\mu$$

(b) Solve the equations to find the values of λ and μ.

(c) What is the position vector of the point of intersection?

(d) Why are two different parameters, λ and μ, used in the vector equations?

F2 Two lines are given by these vector equations, where λ and μ are scalar parameters.

$$\mathbf{r} = 2\mathbf{j} + \lambda(2\mathbf{i} + \mathbf{j}) \qquad \qquad \mathbf{r} = 3\mathbf{i} + \mathbf{j} + \mu(4\mathbf{i} + 2\mathbf{j})$$

(a) Show that, if the lines intersect, there must be values of λ and μ that satisfy these simultaneous equations.

$$2\lambda = 3 + 4\mu$$
$$2 + \lambda = 1 + 2\mu$$

(b) What happens if you try to find the values of λ and μ? Can you explain this?

F3 Two lines are given by these vector equations, where λ and μ are scalar parameters.

$$\mathbf{r} = \begin{bmatrix} -3 \\ 5 \end{bmatrix} + \lambda \begin{bmatrix} 2 \\ 1 \end{bmatrix} \qquad \mathbf{r} = \begin{bmatrix} 4 \\ 1 \end{bmatrix} + \mu \begin{bmatrix} 1 \\ -2 \end{bmatrix}$$

Use simultaneous equations to find where these two lines intersect.

D **F4** Two lines are given by these vector equations, where λ and μ are scalar parameters.

$$\mathbf{r} = \mathbf{i} + 7\mathbf{j} + \lambda(\mathbf{i} + 2\mathbf{j}) \qquad \mathbf{r} = -\mathbf{i} + 3\mathbf{j} + \mu(-2\mathbf{i} - 4\mathbf{j})$$

Use simultaneous equations to find where these two lines intersect.

In three dimensions there is a new possible relationship between two straight lines. For example, in the cuboid $ABCDEFGH$, the straight lines through HB and EG are not parallel but they do not intersect.

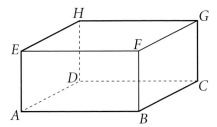

K Two lines that are not parallel but do not intersect are called **skew lines**.

F5 Describe the relationship between these pairs of lines as parallel, intersecting or skew.

(a) AB and HG (b) EF and FG (c) HB and AC

(d) GA and EB (e) HF and DB (f) EF and DB

Two lines are given by the following vector equations.

$$\mathbf{r} = \begin{bmatrix} 1 \\ 2 \\ 4 \end{bmatrix} + \lambda \begin{bmatrix} 2 \\ 3 \\ -1 \end{bmatrix} \qquad \mathbf{r} = \begin{bmatrix} 3 \\ 6 \\ -2 \end{bmatrix} + \mu \begin{bmatrix} 4 \\ 5 \\ 0 \end{bmatrix}$$

The lines are not parallel as the direction vectors $\begin{bmatrix} 2 \\ 3 \\ -1 \end{bmatrix}$ and $\begin{bmatrix} 4 \\ 5 \\ 0 \end{bmatrix}$ are not multiples of each other.

It is not obvious whether the lines intersect or are skew.

D **F6** Can you decide?

Example 6

Two lines are given by the following vector equations.

$$\mathbf{r} = \begin{bmatrix} 2 \\ 3 \\ 5 \end{bmatrix} + t \begin{bmatrix} 4 \\ -1 \\ 3 \end{bmatrix} \qquad \mathbf{r} = \begin{bmatrix} 4 \\ 7 \\ 2 \end{bmatrix} + s \begin{bmatrix} 2 \\ -2 \\ 3 \end{bmatrix}$$

Show that the two lines intersect and find the point of intersection.

Solution

If the two lines intersect then there must be values of t and s such that

$$\begin{bmatrix} 2 \\ 3 \\ 5 \end{bmatrix} + t \begin{bmatrix} 4 \\ -1 \\ 3 \end{bmatrix} = \begin{bmatrix} 4 \\ 7 \\ 2 \end{bmatrix} + s \begin{bmatrix} 2 \\ -2 \\ 3 \end{bmatrix}$$

This gives the equations
$$2 + 4t = 4 + 2s$$
$$3 - t = 7 - 2s$$
$$5 + 3t = 2 + 3s$$

Adding the first two equations gives $5 + 3t = 11 \Rightarrow t = 2$ and substitution into the first or second equation gives $s = 3$.

When $t = 2$, $5 + 3t = 11$; when $s = 3$, $2 + 3s = 11$ too; so the third equation is satisfied.

Hence there is a point that is common to both lines.

Substituting $t = 2$ into the first vector equation gives the point with coordinates $(10, 1, 11)$.

Example 7

Two lines are given by the following vector equations.

$$\mathbf{r} = 2\mathbf{i} + 3\mathbf{j} + 6\mathbf{k} + t(4\mathbf{i} - \mathbf{j} + 5\mathbf{k}) \text{ and } \mathbf{r} = 4\mathbf{i} + 7\mathbf{j} + 8\mathbf{k} + s(2\mathbf{i} - 2\mathbf{j} + \mathbf{k})$$

Show that these are skew lines.

Solution

If the lines do intersect then there must be values of t and s such that

$$2\mathbf{i} + 3\mathbf{j} + 6\mathbf{k} + t(4\mathbf{i} - \mathbf{j} + 5\mathbf{k}) = 4\mathbf{i} + 7\mathbf{j} + 8\mathbf{k} + s(2\mathbf{i} - 2\mathbf{j} + \mathbf{k})$$

giving $\quad (2 + 4t)\mathbf{i} + (3 - t)\mathbf{j} + (6 + 5t)\mathbf{k} = (4 + 2s)\mathbf{i} + (7 - 2s)\mathbf{j} + (8 + s)\mathbf{k}$

This gives the equations
$$2 + 4t = 4 + 2s$$
$$3 - t = 7 - 2s$$
$$6 + 5t = 8 + s$$

The first two equations are the same as those in example 6 so they give $t = 2$ and $s = 3$.

When $t = 2$, $6 + 5t = 16$; when $s = 3$, $8 + s = 11$; so the third equation is not satisfied.

So there is no point common to the two lines.

The directions $4\mathbf{i} - \mathbf{j} + 5\mathbf{k}$ and $2\mathbf{i} - 2\mathbf{j} + \mathbf{k}$ are not multiples of one another so the lines are not parallel.

Hence the lines are skew.

Exercise F (answers p 194)

1 Show that lines $\mathbf{r} = \begin{bmatrix} 5 \\ 4 \\ 3 \end{bmatrix} + \lambda \begin{bmatrix} 1 \\ 0 \\ -3 \end{bmatrix}$ and $\mathbf{r} = \begin{bmatrix} 7 \\ 10 \\ 9 \end{bmatrix} + \mu \begin{bmatrix} 0 \\ 2 \\ 4 \end{bmatrix}$ intersect and

find the point of intersection.

2 Show that lines $\mathbf{r} = \begin{bmatrix} 2 \\ 0 \\ -1 \end{bmatrix} + t \begin{bmatrix} 0 \\ 2 \\ 3 \end{bmatrix}$ and $\mathbf{r} = \begin{bmatrix} -3 \\ 0 \\ 2 \end{bmatrix} + s \begin{bmatrix} 5 \\ 4 \\ -1 \end{bmatrix}$ are skew lines.

3 Show that lines $\mathbf{r} = 4\mathbf{i} + 3\mathbf{j} + \lambda(-\mathbf{i} - \mathbf{j} + \mathbf{k})$ and $\mathbf{r} = 4\mathbf{i} - 3\mathbf{j} + \mu(-\mathbf{i} + \mathbf{j} + \mathbf{k})$ intersect and find the point of intersection.

4 Find whether or not the following pairs of lines intersect.
If they intersect, find the coordinates of the common point; if they do not, find whether they are parallel or skew lines.

(a) $\mathbf{r} = \begin{bmatrix} 2 \\ 3 \\ 5 \end{bmatrix} + \lambda \begin{bmatrix} 1 \\ 0 \\ 2 \end{bmatrix}$, $\mathbf{r} = \begin{bmatrix} 5 \\ 0 \\ 4 \end{bmatrix} + \mu \begin{bmatrix} 0 \\ 3 \\ 7 \end{bmatrix}$

(b) $\mathbf{r} = \begin{bmatrix} 4 \\ 5 \\ 2 \end{bmatrix} + \lambda \begin{bmatrix} 1 \\ 2 \\ 3 \end{bmatrix}$, $\mathbf{r} = \begin{bmatrix} 1 \\ -3 \\ 4 \end{bmatrix} + \mu \begin{bmatrix} 2 \\ 4 \\ 6 \end{bmatrix}$

(c) $\mathbf{r} = 2\mathbf{j} - 3\mathbf{k} + \lambda(6\mathbf{i} - \mathbf{j} - \mathbf{k})$, $\mathbf{r} = -3\mathbf{i} + 6\mathbf{j} - 4\mathbf{k} + \mu(2\mathbf{i} + \mathbf{j} - \mathbf{k})$

(d) $\mathbf{r} = 2\mathbf{i} + \mathbf{j} + 3\mathbf{k} + \lambda(4\mathbf{i} + 2\mathbf{j} - 5\mathbf{k})$, $\mathbf{r} = -3\mathbf{i} + 6\mathbf{j} - 8\mathbf{k} + \mu(3\mathbf{i} - \mathbf{j} + 2\mathbf{k})$

(e) $\mathbf{r} = 4\mathbf{i} + 3\mathbf{j} + \mathbf{k} + \lambda(5\mathbf{i} - 2\mathbf{j} + 4\mathbf{k})$, $\mathbf{r} = 2\mathbf{i} + \mathbf{j} + 6\mathbf{k} + \mu(3\mathbf{i} + \mathbf{j} + 2\mathbf{k})$

5 The following lines intersect.

$$\mathbf{r} = \begin{bmatrix} -4 \\ 3 \\ 2 \end{bmatrix} + \lambda \begin{bmatrix} 2 \\ -1 \\ 0 \end{bmatrix}, \quad \mathbf{r} = \begin{bmatrix} 2 \\ 6 \\ 6 \end{bmatrix} + \mu \begin{bmatrix} 0 \\ 3 \\ p \end{bmatrix}$$

(a) Find p.

(b) Find the coordinates of the point of intersection.

6 The following lines intersect.

$$\mathbf{r} = 2\mathbf{j} - \mathbf{k} + \lambda(\mathbf{i} + 2\mathbf{j} - \mathbf{k})$$
$$\mathbf{r} = 3\mathbf{i} + 4\mathbf{j} + a\mathbf{k} + \mu(5\mathbf{i} + 2\mathbf{j} + 9\mathbf{k})$$

Find the value of the constant a and the coordinates of the point of intersection.

*7 The following lines intersect.

$$\mathbf{r} = \alpha\mathbf{i} + 2\alpha\mathbf{j} + 6\mathbf{k} + p(2\mathbf{i} + \mathbf{j} - \mathbf{k})$$
$$\mathbf{r} = 2\mathbf{i} + 10\mathbf{j} + \mathbf{k} + q(\mathbf{i} - 2\mathbf{j} + \mathbf{k})$$

(a) Find α.

(b) Find the coordinates of the point of intersection.

G Angles and the scalar product (answers p 194)

The angle between two vectors is defined as the angle θ that is formed when the vectors are placed 'tail to tail' or 'head to head' so that $0 \le \theta \le 180°$.

For example, to find the angle between these two vectors **a** and **b** …

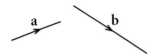

put them …   'tail to tail' …   or 'head to head' …

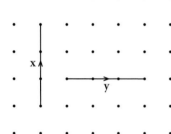   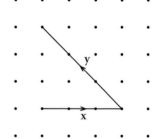

… and the angle θ is the angle between them.

G1 What is the angle between each pair of vectors **x** and **y**?

(a)   (b)   (c)

G2 Vectors **a** and **b** are given by $\mathbf{a} = \begin{bmatrix} -1 \\ 7 \end{bmatrix}$ and $\mathbf{b} = \begin{bmatrix} 4 \\ 3 \end{bmatrix}$.

(a) Draw a diagram that shows **a** and **b** and the angle θ between them.

(b) Find $|\mathbf{a}|$ and $|\mathbf{b}|$.

(c) Find $|\mathbf{b} - \mathbf{a}|$.

(d) Use the cosine rule to find θ, correct to one decimal place.

G3 Vectors **p** and **q** are given by $\mathbf{p} = \mathbf{i} + 2\mathbf{j} - 2\mathbf{k}$ and $\mathbf{q} = 4\mathbf{j} - 5\mathbf{k}$.

(a) Find $|\mathbf{p}|$ and $|\mathbf{q}|$.

(b) Find $|\mathbf{q} - \mathbf{p}|$.

(c) Use the cosine rule to find the angle between **p** and **q**.

There is a simpler way to find the angle between two vectors.

Consider the two vectors $\mathbf{a} = \begin{bmatrix} a_1 \\ a_2 \\ a_3 \end{bmatrix}$ and $\mathbf{b} = \begin{bmatrix} b_1 \\ b_2 \\ b_3 \end{bmatrix}$ and let θ be the angle between them.

Let $\mathbf{c} = \mathbf{b} - \mathbf{a} = \begin{bmatrix} b_1 - a_1 \\ b_2 - a_2 \\ b_3 - a_3 \end{bmatrix}$ so that the vectors $\mathbf{a}$, $\mathbf{b}$ and $\mathbf{c}$ form a triangle.

So $|\mathbf{c}|^2 = (b_1 - a_1)^2 + (b_2 - a_2)^2 + (b_3 - a_3)^2$
$$= b_1^2 + a_1^2 - 2a_1b_1 + b_2^2 + a_2^2 - 2a_2b_2 + b_3^2 + a_3^2 - 2a_3b_3$$
$$= a_1^2 + a_2^2 + a_3^2 + b_1^2 + b_2^2 + b_3^2 - 2(a_1b_1 + a_2b_2 + a_3b_3)$$
$$= |\mathbf{a}|^2 + |\mathbf{b}|^2 - 2(a_1b_1 + a_2b_2 + a_3b_3)$$

Now from the cosine rule we have
$$|\mathbf{c}|^2 = |\mathbf{a}|^2 + |\mathbf{b}|^2 - 2|\mathbf{a}||\mathbf{b}|\cos\theta$$

Comparing the two results, we have $|\mathbf{a}||\mathbf{b}|\cos\theta = a_1b_1 + a_2b_2 + a_3b_3$.

K It can be rearranged to give $\cos\theta = \dfrac{a_1b_1 + a_2b_2 + a_3b_3}{|\mathbf{a}||\mathbf{b}|}$.

G4 Vectors $\mathbf{a}$ and $\mathbf{b}$ are given by $\mathbf{a} = \begin{bmatrix} 3 \\ 2 \\ 1 \end{bmatrix}$ and $\mathbf{b} = \begin{bmatrix} 3 \\ -5 \\ -4 \end{bmatrix}$ and θ is the angle between them.

   (a) Evaluate $a_1b_1 + a_2b_2 + a_3b_3$.
   (b) Find $|\mathbf{a}|$ and $|\mathbf{b}|$.
   (c) Find the value of $\cos\theta$ and hence find the value of θ.

K The **scalar product** of two vectors $\mathbf{a} = \begin{bmatrix} a_1 \\ a_2 \\ a_3 \end{bmatrix}$ and $\mathbf{b} = \begin{bmatrix} b_1 \\ b_2 \\ b_3 \end{bmatrix}$ is defined as the value of

$|\mathbf{a}||\mathbf{b}|\cos\theta$ (where θ is the angle between them) or the value of $a_1b_1 + a_2b_2 + a_3b_3$.

The scalar product of $\mathbf{a}$ and $\mathbf{b}$ is written as $\mathbf{a.b}$, known as 'a dot b'.
It is sometimes called the 'dot product'.

So $\mathbf{a.b} = a_1b_1 + a_2b_2 + a_3b_3$

and $\mathbf{a.b} = |\mathbf{a}||\mathbf{b}|\cos\theta$

are two different ways of finding the scalar product $\mathbf{a.b}$.

G5 Vectors $\mathbf{a}$ and $\mathbf{b}$ are given by $\mathbf{a} = 2\mathbf{i} + \mathbf{j} + 4\mathbf{k}$ and $\mathbf{b} = \mathbf{i} - \mathbf{j} + \mathbf{k}$.
   Show that $\mathbf{a.b} = 5$.

G6 Points A and B lie on a circle, centre O, with a radius of $5\,\text{cm}$ such that angle $AOB = 35°$.
   Find $\overrightarrow{OA}.\overrightarrow{OB}$ correct to two decimal places.

G7 Vectors **a** and **b** are given by $\mathbf{a} = \begin{bmatrix} 3 \\ -1 \\ 1 \end{bmatrix}$ and $\mathbf{b} = \begin{bmatrix} 2 \\ -5 \\ -11 \end{bmatrix}$ and θ is the angle between them.

    (a) Evaluate **a.b**.

    (b) What can you deduce about the value of $\cos\theta$?

    (c) What does this tell you about the vectors **a** and **b**?

Vectors such as $0\mathbf{i} + 0\mathbf{j}$ and $0\mathbf{i} + 0\mathbf{j} + 0\mathbf{k}$ are called zero vectors.

If $\mathbf{a.b} = 0$ then $|\mathbf{a}|\,|\mathbf{b}|\cos\theta = 0$.

If **a** and **b** are non-zero vectors, then this implies that $\cos\theta = 0$ and so $\theta = 90°$.

K   For any two non-zero vectors **a** and **b**, if $\mathbf{a.b} = 0$ then **a** and **b** are perpendicular.

G8 Show that vectors $2\mathbf{i} - 3\mathbf{j} + \mathbf{k}$ and $4\mathbf{i} + \mathbf{j} - 5\mathbf{k}$ are perpendicular.

Example 8

Find the angle between the vectors $2\mathbf{i} + 6\mathbf{j} - \mathbf{k}$ and $\mathbf{i} - 3\mathbf{j} + 3\mathbf{k}$.

Solution

The scalar product of the vectors is $(2\mathbf{i} + 6\mathbf{j} - \mathbf{k}).(\mathbf{i} - 3\mathbf{j} + 3\mathbf{k})$

$$= (2\times1) + (6\times-3) + (-1\times3)$$
$$= -19$$

The magnitude of the vector $2\mathbf{i} + 6\mathbf{j} - \mathbf{k}$ is $\sqrt{2^2 + 6^2 + (-1)^2} = \sqrt{41}$.

The magnitude of the vector $\mathbf{i} - 3\mathbf{j} + 3\mathbf{k}$ is $\sqrt{1^2 + (-3)^2 + 3^2} = \sqrt{19}$.

So the formula gives $\cos\theta = \dfrac{-19}{\sqrt{41}\sqrt{19}} = -0.680\,74\ldots$

So the angle between the vectors is $132.9°$ (correct to 1 d.p.).

Exercise G (answers p 195)

1 $\mathbf{a} = \begin{bmatrix} 2 \\ 2 \\ 1 \end{bmatrix}$, $\mathbf{b} = \begin{bmatrix} 1 \\ 0 \\ -2 \end{bmatrix}$, $\mathbf{c} = \begin{bmatrix} 8 \\ 0 \\ 4 \end{bmatrix}$

    (a) Calculate the scalar products **a.b**, **b.c** and **a.c**.

    (b) Which pairs of these vectors are perpendicular?

2 Find the angles between these pairs of vectors.

    (a) $\begin{bmatrix} 10 \\ 1 \end{bmatrix}, \begin{bmatrix} 2 \\ -5 \end{bmatrix}$ (b) $\begin{bmatrix} 12 \\ 1 \\ -12 \end{bmatrix}, \begin{bmatrix} -8 \\ 4 \\ 1 \end{bmatrix}$ (c) $\begin{bmatrix} 4 \\ -1 \\ -8 \end{bmatrix}, \begin{bmatrix} -7 \\ 4 \\ -4 \end{bmatrix}$

3 Find the angle between the vectors $\mathbf{i} + 8\mathbf{j} + 5\mathbf{k}$ and $-3\mathbf{i} - \mathbf{j} + 4\mathbf{k}$.

4 Triangle PQR is defined by points $P(5, -3, 1)$, $Q(-2, 1, 5)$ and $R(9, 5, 0)$.
Find the angles of the triangle.

5 Show that, for any vector $\mathbf{a} = a_1\mathbf{i} + a_2\mathbf{j} + a_3\mathbf{k}$, $\mathbf{a}.\mathbf{a} = |\mathbf{a}|^2$.

6 Vectors $\mathbf{a}$ and $\mathbf{b}$ are such that $|\mathbf{a}| = 3$, $|\mathbf{b}| = 4$ and $\mathbf{a}.\mathbf{b} = 12$.
What can you say about vectors $\mathbf{a}$ and $\mathbf{b}$?

7 Show that $(0, 0, 0)$, $(4, -2, 5)$, $(3, 6, 0)$ and $(7, 4, 5)$ are four vertices of a square.

H The angle between two straight lines (answers p 195)

In two dimensions, the acute angle θ between two
intersecting straight lines is clearly defined as shown.

The obtuse angle between them is $180 - \theta$.

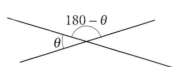

Where two straight line segments in two dimensions
do not intersect, the acute angle between them is defined
as the angle between the extended lines as shown.

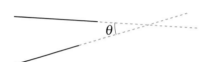

In three dimensions, defining the angles between
two straight lines is more complicated.

Where two straight lines intersect, an angle
between them is clearly defined.

For example, in the cuboid $ABCDEFGH$ the
lines AG and HB intersect and one of the
angles between them is θ as shown.

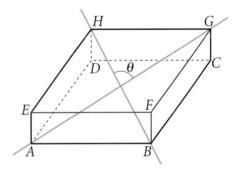

H1 (a) Do the lines AC and HB intersect?

    (b) How would you define an angle between these lines?

H2 What is the angle between the lines through EF and GC?

To find the acute (or obtuse) angle between two skew lines, translate the lines
until they intersect and then find the acute (or obtuse) angle between them.

H3 $PQRSTUVW$ is a cube.
Find the acute angles between these pairs of lines.

  (a) WS and UV (b) TS and QR

  (c) WP and UR (d) TS and PR

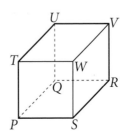

K An angle between two straight lines can be found by finding the angle between any two direction vectors for the lines.

Example 9

ABCDEFGH is a cuboid.
The coordinates of *A*, *F*, *G* and *H* are shown.

Find the acute angle between the line through *A* and *G* and the line through *H* and *F*.

Solution

$\overrightarrow{AG}$ is a direction vector for the line through *A* and *G*.

$\overrightarrow{HF}$ is a direction vector for the line through *H* and *F*.

Let θ be the angle between the direction vectors.

$\overrightarrow{AG} = 4\mathbf{i} + 3\mathbf{j} + 2\mathbf{k}$ and $\overrightarrow{HF} = (4\mathbf{i} + 2\mathbf{k}) - (3\mathbf{j} + 2\mathbf{k}) = 4\mathbf{i} - 3\mathbf{j}$

$\overrightarrow{AG}.\overrightarrow{HF} = (4\mathbf{i} + 3\mathbf{j} + 2\mathbf{k}).(4\mathbf{i} - 3\mathbf{j}) = (4 \times 4) + (3 \times -3) + (2 \times 0) = 7$

The magnitude of $\overrightarrow{AG}$ is $\sqrt{4^2 + 3^2 + 2^2} = \sqrt{29}$

The magnitude of $\overrightarrow{HF}$ is $\sqrt{4^2 + (-3)^2 + 0^2} = \sqrt{25} = 5$

So $\cos\theta = \dfrac{7}{5\sqrt{29}} = 0.259\,97...$

So the angle between these direction vectors is 74.9° (correct to 1 d.p.).

Hence the acute angle between the lines is 74.9°.

D **H4** The equations of two lines l_1 and l_2 are given by

$$l_1: \mathbf{r} = \begin{bmatrix} -2 \\ -11 \\ 17 \end{bmatrix} + \lambda \begin{bmatrix} 3 \\ 5 \\ -1 \end{bmatrix} \qquad l_2: \mathbf{r} = \begin{bmatrix} 9 \\ 7 \\ 0 \end{bmatrix} + \mu \begin{bmatrix} 1 \\ 2 \\ 13 \end{bmatrix}$$

How can you show that these lines are perpendicular?

Example 10

The equations of two lines l_1 and l_2 are given by

$$l_1: \mathbf{r} = \begin{bmatrix} 1 \\ 0 \\ 2 \end{bmatrix} + \lambda \begin{bmatrix} 1 \\ 2 \\ 5 \end{bmatrix} \qquad l_2: \mathbf{r} = \begin{bmatrix} -1 \\ 7 \\ 6 \end{bmatrix} + \mu \begin{bmatrix} 3 \\ -5 \\ 1 \end{bmatrix}$$

Find the acute angle between the two lines.

Solution

The scalar product of the direction vectors is $\begin{bmatrix} 1 \\ 2 \\ 5 \end{bmatrix} \cdot \begin{bmatrix} 3 \\ -5 \\ 1 \end{bmatrix} = (1 \times 3) + (2 \times -5) + (5 \times 1)$

$$= 3 - 10 + 5$$
$$= -2$$

The magnitude of the direction vector for l_1 is $\sqrt{1^2 + 2^2 + 5^2} = \sqrt{30}$

The magnitude of the direction vector for l_2 is $\sqrt{3^2 + (-5)^2 + 1^2} = \sqrt{35}$

Let θ be the angle between the direction vectors.

So $\cos\theta = \dfrac{-2}{\sqrt{30}\sqrt{35}} = -0.0617\ldots$

So the angle between these direction vectors is 93.5° (correct to 1 d.p.)

Hence the acute angle between the lines is $180 - 93.5° = 86.5°$ (correct to 1 d.p.)

Example 11

The equations of two lines l_1 and l_2 are given by

$\quad l_1: \mathbf{r} = 2\mathbf{i} - \mathbf{j} + 6\mathbf{k} + \lambda(-2\mathbf{i} - 4\mathbf{j} + 5\mathbf{k})$ $\qquad l_2: \mathbf{r} = 6\mathbf{i} + \mathbf{j} - \mathbf{k} + \lambda(-11\mathbf{i} + 3\mathbf{j} - 2\mathbf{k})$

Show that the lines are perpendicular.

Solution

The scalar product of the direction vectors is $\quad (-2\mathbf{i} - 4\mathbf{j} + 5\mathbf{k}).(-11\mathbf{i} + 3\mathbf{j} - 2\mathbf{k})$

$$= (-2\times-11) + (-4\times3) + (5\times-2)$$
$$= 22 - 12 - 10$$
$$= 0$$

Hence the lines are perpendicular.

Exercise H (answers p 195)

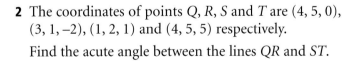

1 *ABCDEF* is a prism where *ABE* and *CDF* are equilateral triangles and each other face is a square.

Find the acute angles between these pairs of lines.

(a) *AB* and *FC*   (b) *AE* and *FC*

(c) *DB* and *EF*   (d) *AD* and *FC*

2 The coordinates of points Q, R, S and T are $(4, 5, 0)$, $(3, 1, -2)$, $(1, 2, 1)$ and $(4, 5, 5)$ respectively.

Find the acute angle between the lines QR and ST.

3 Show that the following two lines are perpendicular.

$\quad l_1: \mathbf{r} = 2\mathbf{i} + 3\mathbf{j} - \mathbf{k} + \lambda(3\mathbf{i} - 6\mathbf{j})$ $\quad l_2: \mathbf{r} = -2\mathbf{i} - 7\mathbf{j} + 6\mathbf{k} + \mu(2\mathbf{i} + \mathbf{j} - 3\mathbf{k})$

4 Find the acute angle between the lines l_1 and l_2 given by

$$l_1: \mathbf{r} = \begin{bmatrix} -3 \\ 0 \\ 2 \end{bmatrix} + \lambda \begin{bmatrix} 2 \\ 3 \\ -1 \end{bmatrix} \qquad l_2: \mathbf{r} = \begin{bmatrix} 4 \\ 8 \\ 1 \end{bmatrix} + \mu \begin{bmatrix} -5 \\ 2 \\ -1 \end{bmatrix}$$

5 The equations of two lines l_1 and l_2 are given by

$\quad l_1: \mathbf{r} = 7\mathbf{i} + \mathbf{j} + 3\mathbf{k} + \lambda(3\mathbf{i} - 4\mathbf{j} + 2\mathbf{k})$ $\quad l_2: \mathbf{r} = -6\mathbf{i} - \mathbf{j} + 3\mathbf{k} + \mu(5\mathbf{i} + 3\mathbf{j} - \mathbf{k})$

Find the acute angle between the lines.

❙ Shortest distance (answers p 196)

I1 In this diagram, point A is fixed and
P is any point on the straight line.
θ is the angle between AP and the line.

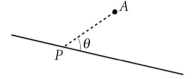

What can you say about θ when P is as close as possible to A?

K The shortest distance from a point to a line is always the distance along the perpendicular from the point to the line.

It is often useful to be able to find this shortest distance and the coordinates of the **foot** of the perpendicular, that is the point where it intersects the line.

The following example in two dimensions illustrates the technique of using the scalar product to find this shortest distance.

Example 12

Find the shortest distance from the point $A\,(4, 7)$ to the line with equation $\mathbf{r} = \begin{bmatrix} 1 \\ 0 \end{bmatrix} + \lambda \begin{bmatrix} 5 \\ 2 \end{bmatrix}$.

Solution

Let P be the point on the line that is closest to A.

As P is on the line then there must be some

value of λ for which $\overrightarrow{OP} = \begin{bmatrix} 1 + 5\lambda \\ 2\lambda \end{bmatrix}$.

$\overrightarrow{AP} = \overrightarrow{OP} - \overrightarrow{OA} = \begin{bmatrix} 1 + 5\lambda \\ 2\lambda \end{bmatrix} - \begin{bmatrix} 4 \\ 7 \end{bmatrix} = \begin{bmatrix} 5\lambda - 3 \\ 2\lambda - 7 \end{bmatrix}$

The direction of the line is given by the vector $\begin{bmatrix} 5 \\ 2 \end{bmatrix}$ and
we know that $\overrightarrow{AP}$ must be perpendicular to this vector.
So the scalar product of these vectors must be 0.

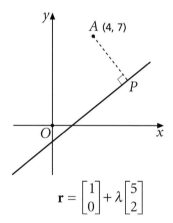

$\begin{bmatrix} 5\lambda - 3 \\ 2\lambda - 7 \end{bmatrix} \cdot \begin{bmatrix} 5 \\ 2 \end{bmatrix} = 0$

A sketch can help you think more clearly about which vectors you need to consider at each stage.

$\Rightarrow \quad 5(5\lambda - 3) + 2(2\lambda - 7) = 0$

$\Rightarrow \quad\quad\quad\quad\quad 29\lambda - 29 = 0$

$\Rightarrow \quad\quad\quad\quad\quad\quad\quad \lambda = 1$

So $\overrightarrow{AP} = \begin{bmatrix} 5 \times 1 - 3 \\ 2 \times 1 - 7 \end{bmatrix} = \begin{bmatrix} 2 \\ -5 \end{bmatrix}$

The shortest distance from A to the line, $|\overrightarrow{AP}|$, is $\sqrt{2^2 + (-5)^2} = \sqrt{29}$.

Example 13

Find the coordinates of the foot of the perpendicular from the point $A\,(10, 9, 11)$ to the line with equation $\mathbf{r} = \mathbf{i} + 2\mathbf{k} + \lambda(2\mathbf{i} + 2\mathbf{j} - \mathbf{k})$.

Solution

Let P be the foot of the perpendicular from A to the line.

As P is on the line then there must be some value of λ for which
$\overrightarrow{OP} = (1 + 2\lambda)\mathbf{i} + 2\lambda\mathbf{j} + (2 - \lambda)\mathbf{k}$, where O is the origin $(0, 0, 0)$.

$$\overrightarrow{AP} = \overrightarrow{OP} - \overrightarrow{OA} = ((1 + 2\lambda)\mathbf{i} + 2\lambda\mathbf{j} + (2 - \lambda)\mathbf{k}) - (10\mathbf{i} + 9\mathbf{j} + 11\mathbf{k})$$
$$= (2\lambda - 9)\mathbf{i} + (2\lambda - 9)\mathbf{j} + (-\lambda - 9)\mathbf{k}$$

The direction of the line is given by the vector $2\mathbf{i} + 2\mathbf{j} - \mathbf{k}$ and we know that $\overrightarrow{AP}$ must be perpendicular to this vector. So the scalar product of these vectors must be 0, that is

$$((2\lambda - 9)\mathbf{i} + (2\lambda - 9)\mathbf{j} + (-\lambda - 9)\mathbf{k}).(2\mathbf{i} + 2\mathbf{j} - \mathbf{k}) = 0$$
$$\Rightarrow \qquad 2(2\lambda - 9) + 2(2\lambda - 9) - (-\lambda - 9) = 0$$
$$\Rightarrow \qquad 9\lambda - 27 = 0$$
$$\Rightarrow \qquad \lambda = 3$$

So $\overrightarrow{OP} = (1 + 2\times3)\mathbf{i} + (2\times3)\mathbf{j} + (2 - 3)\mathbf{k} = 7\mathbf{i} + 6\mathbf{j} - \mathbf{k}$ and so the coordinates of P are $(7, 6, -1)$.

Exercise I (answers p 196)

1 O is the origin $(0, 0)$.
The point P lies on the line $\mathbf{r} = \begin{bmatrix} 5 \\ 4 \end{bmatrix} + \lambda \begin{bmatrix} -1 \\ 3 \end{bmatrix}$ such that $\overrightarrow{OP}$ is perpendicular to the line.
Find the coordinates of P.

2 A line has vector equation $\mathbf{r} = 6\mathbf{i} + \mathbf{j} + t(-2\mathbf{i} + \mathbf{j})$.
The point X has coordinates $(-3, 3)$.
Find the coordinates of the foot of the perpendicular from X to the line.

3 Find the shortest distance from the point $(11, 7)$ to the line $\mathbf{r} = 2\mathbf{i} + \lambda(\mathbf{i} + 3\mathbf{j})$.

4 O is the origin $(0, 0, 0)$.
The point M lies on the line $\mathbf{r} = \begin{bmatrix} 5 \\ 0 \\ -1 \end{bmatrix} + \lambda \begin{bmatrix} 2 \\ -1 \\ 3 \end{bmatrix}$ such that $\overrightarrow{OM}$ is perpendicular to the line.
Find the coordinates of M.

5 The point Q has coordinates $(9, 3, 12)$.
The point P lies on the line $\mathbf{r} = 3\mathbf{i} + \mathbf{k} + \mu(3\mathbf{j} + \mathbf{k})$ such that $\overrightarrow{QP}$ is perpendicular to the line. Find the length of $\overrightarrow{QP}$.

6 A line has vector equation $\mathbf{r} = -\mathbf{i} + \mathbf{j} + 2\mathbf{k} + t(2\mathbf{i} + \mathbf{k})$.
The point A has coordinates $(2, -3, 1)$.
Find the coordinates of the foot of the perpendicular from A to the line.

7 The points $A(2, 0, -3)$, $B(5, 3, 3)$ and $C(3, 1, 2)$ are three vertices of a triangle.

(a) Find a vector equation for the line joining A and B.

(b) Find the length AB.

(c) Find the shortest distance from C to AB, and hence find the area of the triangle.

Key points

- A vector is a quantity that has both magnitude and direction.
  Vectors with the same magnitude and direction are equal. (pp 112–113)

- The modulus of a vector is its magnitude.
  The modulus of the vector **a** is written $|\mathbf{a}|$. (p 113)

- Any vector parallel to the vector **a** may be written as $\lambda\mathbf{a}$, where λ is a non-zero number, and is sometimes called a scalar multiple of **a**.
  In particular, $-\mathbf{a}$ has the same magnitude but is in the opposite direction to **a**. (p 113–114)

- Vectors can be added and subtracted using the 'triangle law'.

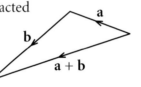 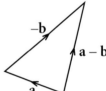

(pp 113–114)

- Vectors can be written in column vector form such as $\begin{bmatrix} 3 \\ -7 \\ 1 \end{bmatrix}$. ← *x*-component, ← *y*-component, ← *z*-component (p 116)

- A unit vector is a vector with a magnitude (or modulus) of 1.
  The vectors **i**, **j** and **k** are unit vectors in the direction of the *x*-, *y*- and *z*-axes respectively.
  As column vectors: $\mathbf{i} = \begin{bmatrix} 1 \\ 0 \\ 0 \end{bmatrix}$, $\mathbf{j} = \begin{bmatrix} 0 \\ 1 \\ 0 \end{bmatrix}$, $\mathbf{k} = \begin{bmatrix} 0 \\ 0 \\ 1 \end{bmatrix}$
  Vectors can be written as linear combinations of these unit vectors.
  For example, $\begin{bmatrix} 3 \\ -7 \\ 1 \end{bmatrix} = 3\mathbf{i} - 7\mathbf{j} + \mathbf{k}$. (pp 118, 121)

- The magnitude (or modulus) of the vector $x\mathbf{i} + y\mathbf{j} + z\mathbf{k}$ is $\sqrt{x^2 + y^2 + z^2}$.
  The distance between two points (x_1, y_1, z_1) and (x_2, y_2, z_2) is
  $$\sqrt{(x_2 - x_1)^2 + (y_2 - y_1)^2 + (z_2 - z_1)^2}.$$ (pp 117, 123)

- For every point P there is associated a unique vector OP (where O is a fixed origin) which is called the **position vector** of the point P: it is sometimes written as **p**.

  The point with coordinates (x, y, z) has position vector $\begin{bmatrix} x \\ y \\ z \end{bmatrix}$ or $x\mathbf{i} + y\mathbf{j} + z\mathbf{k}$. (p 126)

- For two points A and B with position vectors $\overrightarrow{OA}$ and $\overrightarrow{OB}$ respectively, the vector $\overrightarrow{AB}$ is given by $\overrightarrow{OB} - \overrightarrow{OA}$. (p 127)

- The general form of a vector equation of a line is $\mathbf{r} = \mathbf{p} + t\mathbf{d}$ or

  the position vector of *any* point on the line, sometimes written as $\begin{bmatrix} x \\ y \\ z \end{bmatrix}$

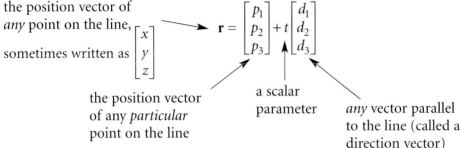

  $$\mathbf{r} = \begin{bmatrix} p_1 \\ p_2 \\ p_3 \end{bmatrix} + t \begin{bmatrix} d_1 \\ d_2 \\ d_3 \end{bmatrix}$$

  the position vector of any *particular* point on the line

  a scalar parameter

  any vector parallel to the line (called a direction vector)

  Using $\mathbf{i}, \mathbf{j}, \mathbf{k}$ notation this equation is
  $$\mathbf{r} = (p_1\mathbf{i} + p_2\mathbf{j} + p_3\mathbf{k}) + t(d_1\mathbf{i} + d_2\mathbf{j} + d_3\mathbf{k})$$ (pp 130–133)

- Lines are parallel if their direction vectors are parallel. (p 133)

- In three dimensions, a pair of lines may be parallel, intersecting or skew. Skew lines are not parallel but nor do they intersect. (p 137)

- To show that two non-parallel straight lines intersect, find values of the parameters in their vector equations that produce the same point on each line. If no such values can be found then the lines are skew. (pp 136–138)

- The angle θ between two vectors is defined as the one formed when the vectors are placed 'tail to tail' or 'head to head' so that $0 \leq \theta \leq 180°$.

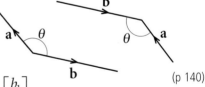

(p 140)

- For two vectors $\mathbf{a} = \begin{bmatrix} a_1 \\ a_2 \\ a_3 \end{bmatrix}$ or $a_1\mathbf{i} + a_2\mathbf{j} + a_3\mathbf{k}$ and $\mathbf{b} = \begin{bmatrix} b_1 \\ b_2 \\ b_3 \end{bmatrix}$ or $b_1\mathbf{i} + b_2\mathbf{j} + b_3\mathbf{k}$,

  the scalar product is written as $\mathbf{a}.\mathbf{b}$ and is defined as
  $$\mathbf{a}.\mathbf{b} = a_1b_1 + a_2b_2 + a_3b_3$$

  or $\quad \mathbf{a}.\mathbf{b} = |\mathbf{a}||\mathbf{b}|\cos\theta$ (where θ is the angle between the vectors)

  Hence the angle between two vectors is given by $\cos\theta = \dfrac{a_1b_1 + a_2b_2 + a_3b_3}{|\mathbf{a}||\mathbf{b}|}$. (p 141)

- Vectors $\mathbf{a}$ and $\mathbf{b}$ are perpendicular if and only if $\mathbf{a}.\mathbf{b} = 0$. You can use this to find the shortest distance from a point to a line. (pp 142, 146)

- One angle between two straight lines is the angle between their direction vectors. (p 144)

Mixed questions (answers p 196)

1 The equations of the lines l_1 and l_2 are given by
$$l_1: \mathbf{r} = 2\mathbf{i} + \mathbf{j} + \lambda(\mathbf{i} + \mathbf{j} - 2\mathbf{k})$$
$$l_2: \mathbf{r} = 5\mathbf{i} + 2\mathbf{j} - \mathbf{k} + \mu(2\mathbf{i} + \mathbf{k})$$
where λ and μ are parameters.

(a) Show that l_1 and l_2 intersect and find the coordinates of B, their point of intersection.

(b) Show that l_1 is perpendicular to l_2.

(c) Show that the point A with coordinates $(1, 0, 2)$ lies on the line l_1.

(d) The point C lies on the line l_2 and has coordinates $(1, p, q)$, where p and q are constants. Find the values of p and q.

(e) Find, in its simplest form, the exact area of the triangle ABC.

2 The points A and B have coordinates $(1, 1, -2)$ and $(2, -3, 0)$ respectively.

(a) Find, in vector form, an equation of the line l_1 that passes through A and B.

The line l_2 passes through the origin and the point C with coordinates $(7, -3, 8)$.

(b) Using a different scalar parameter from the one used in part (a), find a vector equation for l_2.

(c) Show that lines l_1 and l_2 are skew.

(d) Find the shortest distance from C to l_1.

3 Relative to a fixed origin O, points A, B and C have position vectors $-4\mathbf{i} - 3\mathbf{j} + 7\mathbf{k}$, $5\mathbf{i} + 4\mathbf{k}$ and $p\mathbf{i} + \mathbf{j} + q\mathbf{k}$ respectively, where p and q are constants.

(a) Find in vector form, an equation of the line l which passes through A and B.

The point C lies on l.

(b) Find the value of p and the value of q.

(c) Calculate, in degrees, the acute angle between OC and AB.

Point D lies on AB and is such that OD is perpendicular to AB.

(d) Find the position vector of D.

4 (a) Find a vector equation of the line l_1 through the points $A(2, 1, 1)$ and $B(3, 1, 0)$.

(b) The line l_2 has equation $\mathbf{r} = 3\mathbf{j} - \mathbf{k} + \mu(-2\mathbf{i} + \mathbf{j})$.

Show that l_1 and l_2 intersect and find the coordinates of their point of intersection.

(c) Show that the point $C(-8, 7, -1)$ lies on l_2.

(d) Find the coordinates of the point D on l_1 such that CD is perpendicular to l_1.

5 Find the acute angle between the two lines that have vector equations given by
$$\mathbf{r} = \mathbf{i} + 2\mathbf{k} + \lambda(-\mathbf{i} + 5\mathbf{j} + 2\mathbf{k}) \text{ and } \mathbf{r} = 2\mathbf{i} - \mathbf{j} + 5\mathbf{k} + \mu(3\mathbf{i} - 2\mathbf{j} + 6\mathbf{k})$$

Test yourself (answers p 197)

1 The points A, B and C have position vectors $2\mathbf{i} + \mathbf{j} + \mathbf{k}$, $5\mathbf{i} + 7\mathbf{j} + 4\mathbf{k}$ and $\mathbf{i} - \mathbf{j}$ respectively, relative to a fixed origin O.

(a) Prove that the points A, B and C lie on a straight line l.

The point D has position vector $2\mathbf{i} + \mathbf{j} - 3\mathbf{k}$.

(b) Find the cosine of the acute angle between l and the line OD.

The point E has position vector $-3\mathbf{j} - \mathbf{k}$.

(c) Prove that E lies on l and that OE is perpendicular to OD.   Edexcel

2 The line l passes through the points $(3, 1, -2)$ and $(-5, 0, 1)$.

(a) Find a vector equation for l.

(b) Show that the shortest distance between the point $Q(6, 5, -5)$ and l is $\frac{1}{2}\sqrt{62}$.

3 PQR is a triangle where point P has position vector $5\mathbf{i} + 2\mathbf{j} - 3\mathbf{k}$, point Q has position vector $3\mathbf{i} + \mathbf{j} - \mathbf{k}$ and point R has position vector $6\mathbf{i} - 5\mathbf{j} + \mathbf{k}$.

(a) (i) Find the cosine of angle PQR.

(ii) Hence show that the sine of angle PQR is $\dfrac{5\sqrt{17}}{21}$.

(b) Find the exact value of the area of triangle PQR.

4 The line l_1 has equation $\mathbf{r} = 5\mathbf{i} + \mathbf{j} - \mathbf{k} + t(2\mathbf{i} + \mathbf{j} + 5\mathbf{k})$.

The line l_2 has equation $\mathbf{r} = 13\mathbf{i} - 6\mathbf{j} + 2\mathbf{k} + s(-3\mathbf{i} + 4\mathbf{j} + \mathbf{k})$.

(a) Show that l_1 and l_2 intersect and find the coordinates of their point of intersection.

(b) The point Q has coordinates $(10, 11, 7)$.

Find the coordinates of the foot of the perpendicular from Q to l_1.

5 Relative to a fixed origin O, the vector equations of two lines l_1 and l_2 are

$$l_1: \mathbf{r} = 9\mathbf{i} + 2\mathbf{j} + 4\mathbf{k} + t(-8\mathbf{i} - 3\mathbf{j} + 5\mathbf{k})$$

$$\text{and} \quad l_2: \mathbf{r} = -16\mathbf{i} + \alpha\mathbf{j} + 10\mathbf{k} + s(\mathbf{i} - 4\mathbf{j} + 9\mathbf{k})$$

where α is a constant.

The two lines intersect at the point A.

(a) Find the value of α.

(b) Find the position vector of the point A.

(c) Prove that the acute angle between l_1 and l_2 is $60°$.

Point B lies on l_1 and point C lies on l_2.
The triangle ABC is equilateral with sides of length $14\sqrt{2}$.

(d) Find one of the possible position vectors for the point B and the corresponding position vector for the point C.   Edexcel

Answers

1 Partial fractions

A Adding and subtracting algebraic fractions 1

Exercise A (p 7)

1 (a) $\dfrac{3x+1}{(x-1)(x+3)}$

(b) $\dfrac{3}{(x+2)(x+5)}$

(c) $\dfrac{2-7x}{(1+x)(1-2x)}$

(d) $\dfrac{8}{(x+1)(x+3)(x+5)}$

(e) $\dfrac{3(x+32)}{(5+x)(2+x)(x-4)}$

(f) $\dfrac{19-x^2}{(x-1)(3-x)(x+2)}$

2 (a) $\dfrac{3(3x+1)}{2x(x+1)}$ (b) $\dfrac{1-6x}{15(1-x)}$

(c) $\dfrac{5x-4}{6(x+1)(x-2)}$

3 $\dfrac{2(5x-8)}{2x-3}$

4 (a) (i) $\frac{1}{3}$ (ii) $\frac{1}{21}$ (iii) $\frac{1}{171}$

(b) $f(n) = \dfrac{1}{n-1} - \dfrac{2}{2n-1}$

$= \dfrac{2n-1-2(n-1)}{(n-1)(2n-1)}$

$= \dfrac{2n-1-2n+2}{(n-1)(2n-1)}$

$= \dfrac{1}{(n-1)(2n-1)}$

which is a fraction with a numerator of 1.

5 (a) $\dfrac{A}{x+1} + \dfrac{B}{x-1} \equiv \dfrac{A(x-1)+B(x+1)}{(x+1)(x-1)}$

$\equiv \dfrac{Ax-A+Bx+B}{x^2-1} \equiv \dfrac{(A+B)x+(B-A)}{x^2-1}$

(b) $A+B = 5$ and $B-A = 1$ so $A = 2$ and $B = 3$,

and $\dfrac{5x+1}{x^2-1} \equiv \dfrac{2}{x+1} + \dfrac{3}{x-1}$

B Adding and subtracting algebraic fractions

Exercise B (p 9)

1 (a) $\dfrac{1}{4x}$ (b) $\dfrac{6x-1}{x^2}$ (c) $\dfrac{5x+3}{3x^2}$ (d) $\dfrac{2x-}{6x}$

2 (a) $\dfrac{7}{4(x-1)}$ (b) $\dfrac{x+14}{30(4-x^2)}$ or $\dfrac{-x-14}{30(x^2-4)}$

(c) $\dfrac{3x+2}{12x(x-1)}$

3 (a) $\dfrac{3x-2}{(x-1)^2}$ (b) $\dfrac{x}{(x+2)^2}$

(c) $\dfrac{3x+1}{6(x-1)^2}$ (d) $\dfrac{8x+5}{10(2x+1)^2}$

(e) $\dfrac{x+1}{(3x+2)^2}$ (f) $\dfrac{3x-5}{(3-x)^2}$

4 (a) $\dfrac{1}{x+2} - \dfrac{2}{x+4} + \dfrac{x}{(x+4)^2}$

$\equiv \dfrac{(x+4)^2 - 2(x+2)(x+4) + x(x+2)}{(x+2)(x+4)^2}$

$\equiv \dfrac{x^2+8x+16-2x^2-12x-16+x^2+2x}{(x+2)(x+4)^2}$

$\equiv \dfrac{-2x}{(x+2)(x+4)^2}$

(b) When $x > 0$ then $(x+2)(x+4)^2 > 0$ and $-2x < 0$ so $\dfrac{-2x}{(x+2)(x+4)^2} < 0$.

5 (a) $\dfrac{7x+4}{(2-x)(x+1)^2}$ (b) $\dfrac{7}{(2x+1)(x-3)^2}$

(c) $\dfrac{x^2+12x+14}{6(x+1)(x+2)^2}$ (d) $\dfrac{x(x+2)}{(x-1)^3}$

(e) $\dfrac{x+5}{(x+1)(x+3)^2}$ (f) $\dfrac{x^2+x+34}{3(1-x)(2x+1)^2}$

6 (a) $\dfrac{A}{x+2} + \dfrac{B}{(x+2)^2} \equiv \dfrac{A(x+2)+B}{(x+2)^2}$

$\equiv \dfrac{Ax+(2A+B)}{(x+2)^2}$

(b) $A = 2$ and $2A + B = 3$ so $B = -1$, and

$$\frac{2x+3}{(x+2)^2} \equiv \frac{2}{x+2} - \frac{1}{(x+2)^2}$$

C Partial fractions (p 9)

C1 (a)
$$\frac{A}{x+1} + \frac{B}{x+3} \equiv \frac{A(x+3)+B(x+1)}{(x+1)(x+3)}$$
$$\equiv \frac{Ax+3A+Bx+B}{(x+1)(x+3)} \equiv \frac{(A+B)x+(3A+B)}{(x+1)(x+3)}$$

(b) The denominators are identical so the numerators are too. The numerator can be written as $0x + 4$ so we have
$(A + B)x = 0x \Rightarrow A + B = 0$ and $3A + B = 4$.

(c) $A = 2$ and $B = -2$

(d) $\frac{4}{(x+1)(x+3)} \equiv \frac{2}{x+1} + \frac{-2}{x+3}$

which is equivalent to $\frac{2}{x+1} - \frac{2}{x+3}$

C2 (a)
$$\frac{A}{x-1} + \frac{B}{2x+5} \equiv \frac{A(2x+5)+B(x-1)}{(x-1)(2x+5)}$$
$$\equiv \frac{2Ax+5A+Bx-B}{(x-1)(2x+5)} \equiv \frac{(2A+B)x+(5A-B)}{(x-1)(2x+5)}$$

(b) $\frac{1}{x-1} + \frac{5}{2x+5}$

C3 (a) $\frac{2}{x+2} + \frac{1}{2x-1}$

(b) $\frac{1}{x-2} - \frac{1}{x+3}$

(c) $\frac{2}{x-5} - \frac{1}{x-2}$

(d) $\frac{\frac{1}{2}}{3-2x} + \frac{\frac{1}{2}}{3+2x} = \frac{1}{2(3-2x)} + \frac{1}{2(3+2x)}$

C4 (a) $\frac{2}{x-3} - \frac{2}{x+2}$ **(b)** $\frac{1}{x-5} + \frac{1}{x+3}$

C5 (a) $\frac{1}{x+2} - \frac{1}{x+3}$

(b) $\frac{\frac{1}{2}}{x+3} + \frac{\frac{1}{2}}{x-3} = \frac{1}{2(x+3)} + \frac{1}{2(x-3)}$

(c) $\frac{1}{x+1} + \frac{2}{x-1}$

(d) $\frac{5}{x} - \frac{3}{x+1}$

C6 (a) $\frac{1}{x-1} + \frac{1}{x-2} + \frac{1}{x-3}$

(b) $\frac{8}{x} + \frac{3}{1-x} - \frac{5}{x+4}$

C7
$$\frac{AC-B}{(C-D)(x+C)} + \frac{B-AD}{(C-D)(x+D)}$$
$$\equiv \frac{(x+D)(AC-B)+(x+C)(B-AD)}{(C-D)(x+C)(x+D)}$$
$$\equiv \frac{ACx-Bx+ACD-BD+Bx+BC-ADx-ADC}{(C-D)(x+C)(x+D)}$$
$$\equiv \frac{ACx-ADx-BD+BC}{(C-D)(x+C)(x+D)}$$
$$\equiv \frac{Ax(C-D)+B(C-D)}{(C-D)(x+C)(x+D)}$$
$$\equiv \frac{(Ax+B)(C-D)}{(C-D)(x+C)(x+D)}$$
$$\equiv \frac{Ax+B}{(x+C)(x+D)} \quad \text{as required}$$

Exercise C (p 13)

1 (a) $\frac{3}{x} + \frac{3}{1-x}$ **(b)** $\frac{5}{x} - \frac{2}{x+6}$

(c) $\frac{1}{x+3} + \frac{1}{x+4}$ **(d)** $\frac{2}{x-3} - \frac{1}{x+1}$

(e) $\frac{3}{3x-1} - \frac{2}{2x+1}$ **(f)** $\frac{4}{2-x} + \frac{2}{2x-1}$

(g) $\frac{1}{2(x+1)} - \frac{1}{2(x+5)}$ **(h)** $\frac{1}{4(x-1)} + \frac{1}{4(3x+1)}$

(i) $\frac{7}{11(x+2)} - \frac{2}{11(5x-1)}$

2 (a) $\frac{1}{x-2} - \frac{1}{x+2}$ **(b)** $\frac{1}{3x} + \frac{2}{3(x+3)}$

(c) $\frac{3}{x+1} - \frac{7}{3x+2}$ **(d)** $\frac{1}{3(2x-3)} - \frac{1}{3(2x+3)}$

3 (a) $\frac{1}{x-1} - \frac{9}{x+2} + \frac{13}{x+3}$

(b) $-\frac{3}{x} + \frac{2}{x+1} + \frac{1}{x-2}$

(c) $-\frac{1}{x+1} + \frac{4}{2x+1} - \frac{3}{3x+1}$

(d) $\frac{1}{2(x-1)} - \frac{5}{2(x+1)} + \frac{4}{2x+1}$

4 (a) $f(x) = \dfrac{1}{x-3} - \dfrac{1}{x+2}$

(b) $f(x) = (x-3)^{-1} - (x+2)^{-1}$

so $f'(x) = -(x-3)^{-2} - \left(-(x+2)^{-2}\right)$

$= -(x-3)^{-2} + (x+2)^{-2}$

$= \dfrac{1}{(x+2)^2} - \dfrac{1}{(x-3)^2}$ as required.

5 (a) $f(3) = \frac{9}{10}$, $f(4) = \frac{12}{13}$

(b) (i) $\dfrac{1}{3n-2} - \dfrac{1}{3n+1}$

(ii) Part (i) shows that the nth term in the series can be written as the difference of the two fractions $\dfrac{1}{3n-2}$ and $\dfrac{1}{3n+1}$.

So the first term is $\frac{1}{1} - \frac{1}{4}$, the second term is $\frac{1}{4} - \frac{1}{7}$, the third term is $\frac{1}{7} - \frac{1}{10}$ and so on. Hence the whole series can be written as

$\left(\frac{1}{1} - \frac{1}{4}\right) + \left(\frac{1}{4} - \frac{1}{7}\right) + \left(\frac{1}{7} - \frac{1}{10}\right) + \dots +$

$\left(\dfrac{1}{3n-2} - \dfrac{1}{3n+1}\right)$. The series can be written with the terms grouped as

$\frac{1}{1} + \left(-\frac{1}{4} + \frac{1}{4}\right) + \left(-\frac{1}{7} + \frac{1}{7}\right) + \left(-\frac{1}{10} + \frac{1}{10}\right) + \dots +$

$\left(-\dfrac{1}{3n-2} + \dfrac{1}{3n-2}\right) - \dfrac{1}{3n+1}$ so that the sum of each bracketed pair is 0.

Hence the sum of the whole series is $1 - \dfrac{1}{3n+1}$.

As $f(n) = 1 - \dfrac{1}{3n+1}$ and n is positive then $f(n) < 1$ as required.

D Improper fractions

Exercise D (p 14)

1 (a) $A = 3$, $B = -2$, $C = 5$

(b) $A = 2$, $B = -5$, $C = 4$

(c) $A = 8$, $B = 9$, $C = -25$

(d) $A = 2$, $B = -1$, $C = -3$

2 $A = 2$, $B = 1$, $C = 3$, $D = -1$

3 (a) $f(x) = 3x + \dfrac{1}{x+1} + \dfrac{2}{x-1}$

(b) $f'(x) = 3 - \dfrac{1}{(x+1)^2} - \dfrac{2}{(x-1)^2}$

$= 3 - \left(\dfrac{1}{(x+1)^2} + \dfrac{2}{(x-1)^2}\right)$

Now for all $x \in \mathbb{R}$, $x \neq \pm 1$,

$\left(\dfrac{1}{(x+1)^2} + \dfrac{2}{(x-1)^2}\right) > 0$

so $3 - \left(\dfrac{1}{(x+1)^2} + \dfrac{2}{(x-1)^2}\right) < 3$

so $f'(x) < 3$ as required.

E Further partial fractions (p 14)

E1 (a) $\dfrac{x^2 + x + 1}{x(x^2+1)}$ **(b)** $\dfrac{x+1}{x(x^2+1)}$

E2 (a) $\dfrac{A}{x} + \dfrac{B}{x^2+1} \equiv \dfrac{A(x^2+1) + Bx}{x(x^2+1)}$

$\equiv \dfrac{Ax^2 + Bx + A}{x(x^2+1)}$

So $\dfrac{Ax^2 + Bx + A}{x(x^2+1)} \equiv \dfrac{1-2x}{x(x^2+1)}$

$\Rightarrow A = 0$ (from equating coefficients of x^2) a $A = 1$ (from equating constants) which is a contradiction so no such identity exists.

(b) $A = 1$, $B = -1$, $C = -2$

E3 $\dfrac{x+2}{(x+3)^2}$

E4 $\dfrac{A}{x-1} + \dfrac{B}{x-1} \equiv \dfrac{A+B}{x-1} \equiv \dfrac{(A+B)(x-1)}{(x-1)^2}$

$\equiv \dfrac{(A+B)x - (A+B)}{(x-1)^2}$

So $\dfrac{x}{(x-1)^2} \equiv \dfrac{A}{x-1} + \dfrac{B}{x-1}$

$\Rightarrow A + B = 1$ (from equating coefficients of x) an $A + B = 0$ (from equating constants) which is a contradiction so no such identity exists.

E5 (a) (i) $A = 1, B = 1$

   (ii) $A = 3, B = 1$

(b) (i) $\dfrac{A}{x+C} + \dfrac{B-AC}{(x+C)^2} \equiv \dfrac{A(x+C)+B-AC}{(x+C)^2}$

$\equiv \dfrac{Ax+AC+B-AC}{(x+C)^2} \equiv \dfrac{Ax+B}{(x+C)^2}$

   as required

   (ii) $\dfrac{5}{x+3} + \dfrac{1}{(x+3)^2}$

E6 $\dfrac{7x^2+2x-12}{(2x-5)(x+1)^2}$

Exercise E (p 16)

1 (a) $\dfrac{1}{x+2} + \dfrac{2}{(x+2)^2}$

(b) $\dfrac{2}{x+3} - \dfrac{2}{x+1} + \dfrac{4}{(x+1)^2}$

(c) $\dfrac{1}{x-2} + \dfrac{2}{(x-1)^2}$

(d) $\dfrac{1}{x-1} - \dfrac{1}{x+3} - \dfrac{1}{(x+3)^2}$

(e) $\dfrac{1}{2(x-4)} + \dfrac{1}{2(x+2)} - \dfrac{1}{(x+2)^2}$

(f) $-\dfrac{2}{3(x+4)} + \dfrac{2}{3(x-2)} + \dfrac{3}{(x-2)^2}$

(g) $\dfrac{3}{x-5} + \dfrac{2}{2-x} - \dfrac{2}{(2-x)^2}$

(h) $\dfrac{2}{2x+1} - \dfrac{2}{x+3} + \dfrac{1}{(x+3)^2}$

(i) $\dfrac{1}{3(2x-1)} - \dfrac{2}{3(x-5)} - \dfrac{2}{(x-5)^2}$

(j) $\dfrac{1}{7(x+2)} - \dfrac{3}{7(3x-1)} + \dfrac{1}{(3x-1)^2}$

(k) $\dfrac{3}{x} - \dfrac{12}{4x-1} + \dfrac{16}{(4x-1)^2}$

(l) $\dfrac{14}{5(2x+5)} - \dfrac{2}{5x} + \dfrac{1}{x^2}$

2 (a) $f(x) = \dfrac{2}{x-2} - \dfrac{2}{x+1} - \dfrac{1}{(x+1)^2}$

(b) $f'(x) = -2(x-2)^{-2} + 2(x+1)^{-2} + 2(x+1)^{-3}$

$= 2\left((x+1)^{-3} + (x+1)^{-2} - (x-2)^{-2}\right)$

$= 2\left(\dfrac{1}{(x+1)^3} + \dfrac{1}{(x+1)^2} - \dfrac{1}{(x-2)^2}\right)$

   as required.

3 $\dfrac{(8x+1)(x-2)}{(x-3)(2x-1)^2}$ is a proper fraction and so can be

written as $\dfrac{A}{x-3} + \dfrac{B}{2x-1} + \dfrac{C}{(2x-1)^2}$ for constants

A, B and C.

This gives
$A(2x-1)^2 + B(x-3)(2x-1) + C(x-3)$
$\equiv (8x+1)(x-2)$

Substituting $x = \frac{1}{2}$ gives $-\frac{5}{2}C = -\frac{15}{2}$ so $C = 3$

Substituting $x = 3$ gives $25A = 25$ so $A = 1$

Substituting $x = 0$ gives $A + 3B - 3C = -2$

$\Rightarrow \qquad 3B = -2 - A + 3C = 6$

$\Rightarrow \qquad B = 2$

So $f(x) = \dfrac{1}{x-3} + \dfrac{2}{2x-1} + \dfrac{3}{(2x-1)^2}$

Differentiating gives

$f'(x) = \dfrac{-1}{(x-3)^2} - \dfrac{4}{(2x-1)^2} - \dfrac{12}{(2x-1)^3}$

$= -\left(\dfrac{1}{(x-3)^2} + \dfrac{4}{(2x-1)^2} + \dfrac{12}{(2x-1)^3}\right)$

Now if $x > 3$ then each denominator is greater than 0

so $\left(\dfrac{1}{(x-3)^2} + \dfrac{4}{(2x-1)^2} + \dfrac{12}{(2x-1)^3}\right) > 0$ and so

$-\left(\dfrac{1}{(x-3)^2} + \dfrac{4}{(2x-1)^2} + \dfrac{12}{(2x-1)^3}\right) < 0$

So $f'(x) < 0$ for all values in the domain $x > 3$ and so $f(x)$ is decreasing for all values of x in the domain.

1 $\dfrac{x^2}{(x+1)(x-1)^2}$

2 $-\dfrac{2}{x+3}+\dfrac{3}{2x+1}$

3 $\dfrac{3}{1+5x}+\dfrac{2}{3-x}$

4 $\dfrac{1}{x+2}-\dfrac{2}{x-1}+\dfrac{1}{x-4}$

5 (a) $\text{f}(x)=\dfrac{1}{x+2}+\dfrac{2}{x-1}$

  (b) $\text{f}'(x)=-\dfrac{1}{(x+2)^2}-\dfrac{2}{(x-1)^2}$

  $=-\left(\dfrac{1}{(x+2)^2}+\dfrac{2}{(x-1)^2}\right)$

  Now for all values of x in the domain

  $x\in\mathbb{R},\ x\neq-2,\ x\neq1,\ \left(\dfrac{1}{(x+2)^2}+\dfrac{2}{(x-1)^2}\right)>0$

  so $\text{f}'(x)<0$ for all values of x in the domain.

6 $A=4,\ B=1,\ C=2$

7 $\dfrac{1}{2(2x+3)}-\dfrac{5}{2(2x+3)^2}$

8 $\dfrac{9}{2-3x}-\dfrac{3}{1-x}-\dfrac{2}{(1-x)^2}$

2 Parametric equations

A Coordinates in terms of a third variable
(p 18)

A1 (a)

t	0	1	2	3	4	5	6
x	0	3	6	9	12	15	18
y	0	5	8	9	8	5	0

(b)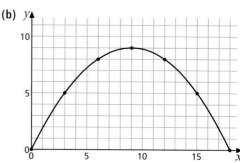

It might represent an object that has been thrown.

A2 (a) $x=2t$ **(b)** $y=t$

A3 (a) The dot accelerates away from the origin, rather than moving at constant speed.

  (b) $x=t^2$ **(c)** $y=\dfrac{t^2}{2}$

A4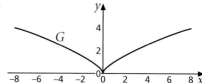

A5 Plots with the following descriptions

  (a) G translated by $\begin{bmatrix}0\\2\end{bmatrix}$

  (b) G translated by $\begin{bmatrix}-1\\0\end{bmatrix}$

  (c) G stretched by factor 3 in the x-direction

  (d) G 'stretched' by factor $\tfrac{1}{2}$ in the y-direction

  (e) G reflected in the x-axis

  (f) G translated by $\begin{bmatrix}4\\-1\end{bmatrix}$

A6 (a) $x=t^2+t+3,\ y=-t$

  (b) $x=2(t^2+t),\ y=1-t$

Exercise A (p 20)

1 $(-\tfrac{1}{3},9),\ (-\tfrac{1}{2},4),\ (-1,1),\ (1,1),\ (\tfrac{1}{2},4),\ (\tfrac{1}{3},9)$

2 (a) $x = t, y = 3t^2$ **(b)** $x = t + 2, y = t^2 + 1$

3 (a) (i) $(2, -3), (3, 0), (4, 1), (5, 0), (6, -3)$

   (ii)

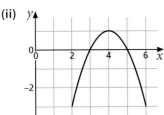

  (b) (i) $(4, -4), (3, 1), (2, 0), (1, -1), (0, 4)$

   (ii)

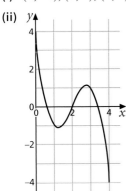

  (c) (i) $(-8, 6), (-1, 2), (0, 0), (1, 0), (8, 2)$

   (ii)

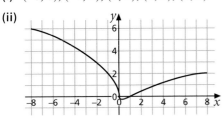

4 (a) 5   **(b)** -24

5 3

6 $-2, \frac{3}{2}$

7 (a) $x = t^2, y = t^2 - t$ **(b)** $x = t^2 - 2, y = t^3 - 3t$

B Converting between parametric and cartesian equations (p 21)

B1 (a)

t	−2	−1.5	−1	−0.5	0	0.5	1	1.5	2
x	−5	−4	−3	−2	−1	0	1	2	3
y	−8	−6	−4	−2	0	2	4	6	8

  (b) $t = \dfrac{y}{4}$

  (c) $\qquad\qquad x = 2\left(\dfrac{y}{4}\right) - 1$

    $\Rightarrow \qquad\qquad x = \dfrac{y}{2} - 1$

    $\Rightarrow \qquad\qquad 2x = y - 2$

    $\Rightarrow 2x - y + 2 = 0$

  (d) A check of the values

  (e) A straight line

B2 (a) $2x - 3y - 8 = 0$ **(b)** $x + 6y - 5 = 0$

B3 (a) $t = \dfrac{x}{3}$

  (b) $\qquad y = 6\left(\dfrac{x}{3}\right) - \left(\dfrac{x}{3}\right)^2$

    $\Rightarrow \ y = 2x - \dfrac{x^2}{9}$

  (c) A check of the table values

B4 These equations or their equivalents:

  (a) $4x - 5y - 15 = 0$ **(b)** $y^2 = 4x$

  (c) $y = \dfrac{x^3}{8}$ **(d)** $3x + y - 9 = 0$

  (e) $y = 1 - \dfrac{x^2}{16}$ **(f)** $y^3 = -x$

B5 (a) $y = x^4$ **(b)** $y = -x^2 + 6x - 6$

  (c) $y = x^3 - 3x^2 + 2$

Exercise B (p 23)

1 These equations or their equivalents:

  (a) $y = 20x - 1$

  (b) $y = \dfrac{16}{x}$

  (c) $x - 9y^2 - 12y = 0$

  (d) $2xy = 1$

  (e) $y = x^2 - 4x + 8$

  (f) $y = 1 - 2x$

  (g) $y = x^4 + x^2$

  (h) $y = 4 - \dfrac{1}{x}$

  (i) $x = y^3 - 6y^2 + 11y - 6$

  (j) $y = -x^3 + 5x^2 - 7x + 3$

  (k) $y = 4x^2 + 10x + 6$

  (l) $x + 4y - 27 = 0$

2 $x + y = 2t^3$

$x - y = \dfrac{6}{t}$

Hence $(x + y)(x - y)^3 = 2t^3 \times \dfrac{216}{t^3} = 432$ as required.

3 (a) $y = 5x^2 - x^4$

(b) $y = \dfrac{1}{x^2} + \dfrac{6}{x} + 9$

(c) $y = \dfrac{1}{x^3} + \dfrac{6}{x^2} + \dfrac{12}{x} + 8$

(d) $\dfrac{y}{x} - 3y - 1 = 0$

(e) $y = 17 - 4x$

(f) $y = \dfrac{x}{1 - x}$

4 $x + y = \dfrac{1}{t} + \dfrac{1}{t(t - 1)}$

$= \dfrac{(t - 1) + 1}{t(t - 1)} = \dfrac{t}{t(t - 1)} = \dfrac{1}{t - 1}$

$\dfrac{y}{x} = \dfrac{1}{t(t - 1)} \div \dfrac{1}{t} = \dfrac{1}{t(t - 1)} \times \dfrac{t}{1} = \dfrac{1}{t - 1}$

Hence $x + y = \dfrac{y}{x}$ is the cartesian equation.

5

$x = \dfrac{1}{2t - 1}$

$\Rightarrow \quad 2t - 1 = \dfrac{1}{x}$

$\Rightarrow \quad 2t = \dfrac{1}{x} + 1$

$\Rightarrow \quad t = \tfrac{1}{2}\left(\dfrac{1}{x} + 1\right)$

Substituting into the y-equation:

$y = \dfrac{\tfrac{1}{2}\left(\dfrac{1}{x} + 1\right)}{\dfrac{1}{x}}$

$\Rightarrow \quad y = \tfrac{1}{2}\left(\dfrac{1}{x} + 1\right) \times \dfrac{x}{1}$

$\Rightarrow \quad y = \tfrac{1}{2} + \tfrac{1}{2}x$, which is a straight line.

Alternatively:

$x = \dfrac{1}{2t - 1}$ and $y = \dfrac{t}{2t - 1}$

Hence $2y - x = \dfrac{2t - 1}{2t - 1}$

$\Rightarrow \quad 2y - x = 1$

C Solving problems (p 24)

C1 (a) $t = 6 - a$

(b) $9 = a(6 - a)$

$\Rightarrow \quad 9 = 6a - a^2$

$\Rightarrow \quad a^2 - 6a + 9 = 0$

(c) $(a - 3)(a - 3) = 0$

So $a = 3$

C2 -3

C3 (a) 2 **(b)** -1

C4 $4, 13$

Exercise C (p 26)

1 4

2 3

3 3

4 $70, -5$

5 $(0, 10), (0, -10)$

6 (a) $(0, 5)$ **(b)** $(0, 6)$

   (c) $(0, -12), (0, -3)$ **(d)** $(0, \tfrac{1}{3})$

7 (a) $(-2.5, 0)$ **(b)** $(3, 0), (2, 0)$

   (c) $(4, 0)$ **(d)** $(7, 0)$

8 $\left(\tfrac{3}{5}, \tfrac{11}{5}\right)$

9 $\left(\tfrac{5}{2}, \tfrac{3}{2}\right)$

10

$\dfrac{t}{t^2 - 2} = 1$

$\Rightarrow \quad t = t^2 - 2$

$\Rightarrow \quad t^2 - 2 - t = 0$

$\Rightarrow \quad (t + 1)(t - 2) = 0$

$\Rightarrow \quad t = -1, 2$

When $t = -1, x = \sqrt{-1 + 1} = 0$

When $t = 2, x = \sqrt{2 + 1} = \sqrt{3}$

So the curve meets $y = 1$ at $(0, 1)$ and $(\sqrt{3}, 1)$.

11 $2(\sqrt{2})^3$ or $4\sqrt{2}$

12 9

13 -2

14 (a) 9 **(b)** $(9, 0), (36, 0)$

15 $(-1, -3), (3, 5)$

16 $(-\sqrt{3}, 2), (0, -1), (\sqrt{3}, 2)$

D Circle and ellipse (p 27)

D1 A check with a graph plotter

D2 These in fact give

  (a) a circle centre $(0, 0)$, radius 3 units

  (b) a circle centre $(0, 0)$, radius 0.5 unit

D3 (a) $x = \cos\theta, y = \sin\theta + 2$

  (b) $x = \cos\theta - 3, y = \sin\theta$

  (c) $x = \cos\theta + 1, y = \sin\theta - 6$

  (d) $x = 2\cos\theta + 5, y = 2\sin\theta + 4$

D4 (a)

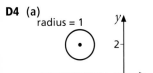

radius = 1

(b)

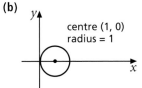

centre (1, 0)
radius = 1

(c)

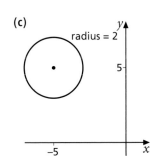

radius = 2

(d)

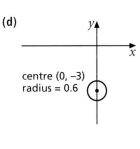

centre (0, –3)
radius = 0.6

D5 (a) Stretch, factor 2, in the y-direction

  (b) Stretch, factor 4, in the x-direction

  (c) 'Stretch', factor 0.6, in the y-direction

  (d) 'Stretch', factor 0.5, in the x-direction and
    stretch, factor 1.2, in the y-direction

D6 (a) $x = 4\cos\theta, y = 2\sin\theta$

  (b) $x = 5\cos\theta, y = 3\sin\theta$

  (c) $x = 2\cos\theta, y = 3\sin\theta$

D7 (a) $x = 2\cos\theta + 3, y = \sin\theta$

  (b) $x = 2\cos\theta, y = \sin\theta - 2$

  (c) $x = 2\cos\theta + 1, y = \sin\theta + 2$

  (d) $x = 2\cos\theta - 1, y = 3\sin\theta - 1$

D8 From the parametric equations,

$$\cos\theta = \frac{x}{a}, \sin\theta = \frac{y}{b}$$

Substituting into $\cos^2\theta + \sin^2\theta = 1$ gives

$$\frac{x^2}{a^2} + \frac{y^2}{b^2} = 1$$

D9 $\dfrac{x^2}{2^2} + \dfrac{y^2}{3^2} = 1$

  $x = 2\cos\theta, y = 3\sin\theta$

D10 (a) $x = 3\cos\theta, y = 6\sin\theta$

  (b) $x = 3\cos\theta, y = \sin\theta$

  (c) $x = \frac{1}{2}\cos\theta, y = 3\sin\theta$

Exercise D (p 30)

1 $(1, \sqrt{3})$

2 (a) $x = 4\cos\theta, y = \sin\theta$ (b) $x = \cos\theta, y = 0.7\sin\theta$

3 $(3, 0), \left(\dfrac{3\sqrt{2}}{2}, 2\sqrt{2}\right), (0, 4), (-3, 0)$

4 $4x^2 + 9y^2 = 1$

5 (a) (i) $x = 2\cos\theta, y = \sin\theta$ (ii) $\dfrac{x^2}{4} + y^2 = 1$

  (b) (i) $x = \frac{1}{2}\cos\theta, y = \sin\theta$ (ii) $4x^2 + y^2 = 1$

  (c) (i) $x = 2\cos\theta, y = 1.5\sin\theta$ (ii) $\dfrac{x^2}{4} + \dfrac{4y^2}{9} = 1$

6 (a) $(3, -2)$ (b) 5 units

  (c) $x = 5\cos\theta + 3, y = 5\sin\theta - 2$

7 (a) (i) Ellipse, centre $(0, 0)$, 4 units wide and
        6 units high

    (ii) $\dfrac{x^2}{4} + \dfrac{y^2}{9} = 1$

  (b) (i) Circle, centre $(0, 1)$, radius 1 unit

    (ii) $x^2 + (y - 1)^2 = 1$

  (c) (i) Circle, centre $(0, 0)$, radius 4 units

    (ii) $\dfrac{x^2}{16} + \dfrac{y^2}{16} = 1$

  (d) (i) Ellipse, centre $(1, 0)$, 2 units wide and
        4 units high

    (ii) $(x - 1)^2 + \dfrac{y^2}{4} = 1$

  (e) (i) Circle, centre $(-3, 0)$, radius 2 units

    (ii) $\dfrac{(x + 3)^2}{4} + \dfrac{y^2}{4} = 1$, or $x^2 + y^2 + 6x + 5 = 0$

  (f) (i) Ellipse, centre $(3, 2)$, 4 units wide and
        8 units high

(ii) $\dfrac{(x-3)^2}{4} + \dfrac{(y-2)^2}{16} = 1$, or

$4x^2 + y^2 - 24x - 4y + 24 = 0$

(g) (i) Circle, centre $(-2, 3)$, radius 5 units

(ii) $\dfrac{(x+2)^2}{25} + \dfrac{(y-3)^2}{25} = 1$, or

$x^2 + y^2 + 4x - 6y - 12 = 0$

(h) (i) Ellipse, centre $(-3, 1)$, 1 unit wide and 4 units high

(ii) $4(x+3)^2 + \dfrac{(y-1)^2}{4} = 1$, or

$16x^2 + y^2 + 96x - 2y + 141 = 0$

8 $x = 6\cos\theta + 3,\ y = 6\sin\theta - 4$

E Using other trigonometric identities

Exercise E (p 32)

1 (a) $y = \dfrac{2}{x}$ **(b)** $y = \dfrac{1}{2x}$ **(c)** $y = \pm\dfrac{1}{\sqrt{x}}$

(d) $y = \dfrac{4}{x-2}$ **(e)** $y = \dfrac{2x^2}{25} - 1$ **(f)** $y = 3 - 6x^2$

2 (a) $\dfrac{y^2}{4} = 1 + x^2$

(b) $y = 2x\sqrt{1 - x^2}$

(c) $\dfrac{x^2}{4} - \dfrac{y^2}{9} - 1 = 0$

(d) $9x^2 + 54x + 90 = y^2$

(e) $-x^2 + 4y^2 + 2x + 16y - 1 = 0$

(f) $x = \dfrac{2(y-1)}{y(2-y)}$

(g) $y^2 = \dfrac{1}{1+x^2}$

(h) $y^2 = 4 + \dfrac{4}{x^2}$

3 $y^2 = x^2(4 - x^2)$

Test yourself (p 34)

1 $(\frac{1}{9}, -9),\ (\frac{1}{4}, -6),\ (1, -3),\ (1, 3),\ (\frac{1}{4}, 6),\ (\frac{1}{9}, 9)$

2 (a) $(-3\frac{1}{3}, 0),\ (0, 10)$

(b) $(15, 0),\ (0, 5),\ (0, 3)$

(c) $(-14, 0),\ (-4, 0),\ (0, 14)$

(d) $(\frac{1}{2}, 0),\ (0, 26)$

3 $(-6, 1\frac{1}{2}),\ (-6, 1\frac{1}{3})$

4 (a) $3x - 4y - 15 = 0$ **(b)** $x + 6y - 5 = 0$

5 (a) $y = \dfrac{x^2}{9}$

(b) $x = 8y^3$

(c) $y = x^6$

(d) $y = -x^2 - 2x + 1$

(e) $y = \dfrac{3}{x}$

(f) $x = 16y^2 - 4y$

(g) $y = x^3 - 6x^2 + 14x - 12$

(h) $y = 3 - \dfrac{1}{2x}$

(i) $x + 3y - 17 = 0$

(j) $x = 3y^2 - y^4$

(k) $x = \dfrac{1}{y^2} + \dfrac{4}{y} + 4$

(l) $y = \dfrac{x}{2x - 1}$

(m) $x + 3y - 10 = 0$

(n) $x = \dfrac{y}{5y - 3}$

6 $(x + y)(x - y)^2 = 32$

7 (a) $x = 2\cos\theta,\ y = 2\sin\theta$

(b) $x = 2\cos\theta,\ y = 3\sin\theta$

(c) $x = 4\cos\theta - 1,\ y = 4\sin\theta - 3$

(d) $x = 3\cos\theta + 4,\ y = \sin\theta + 2$

8 (a) $x = 2\cos\theta,\ y = 4\sin\theta$

(b) $x = 2\cos\theta,\ y = 5\sin\theta$

9 $6\sqrt{3}$

10 (a) $\dfrac{x^2}{9} + \dfrac{y^2}{16} = 1$

(b) $\dfrac{(x-5)^2}{9} + \dfrac{(y+2)^2}{9} = 1$

(c) $y = \dfrac{2}{x}$

(d) $y = \dfrac{1}{(x-1)^2}$

(e) $4x^2 - 9y^2 + 72y - 108 = 0$

(f) $8y^2 + x - 32y + 29 = 0$

(g) $y^2 = \dfrac{x^2}{1 - x^2}$

(h) $y = 2(x-3)\sqrt{1 - (x-3)^2}$

3 The binomial theorem

A Reviewing the binomial theorem for positive integers

Exercise A (p 36)

1 $1 + 30x + 375x^2 + 2500x^3 + 9375x^4 + 18\,750x^5$
$+ 15\,625x^6$

2 $1 - 36x + 594x^2 - 5940x^3 + 40\,095x^4$

3 $16\,384 + 229\,376x + 1\,376\,256x^2$

4 (a) $1 + 4x + 7x^2 + 7x^3$

(b) $256 + 1024x + 1792x^2 + 1792x^3$

5 (a) $729 + 1458x + 1215x^2 + 540x^3 + 135x^4 + 18x^5$
$+ x^6$

(b) 743.70

B Extending the binomial theorem (p 37)

B1 (a) The coefficients for the x^2, x^3 and x^4 terms are

$$\frac{(-2)(-3)}{2!} = \frac{-2 \times -3}{2 \times 1} = \frac{6}{2} = 3$$

$$\frac{(-2)(-3)(-4)}{3!} = \frac{-2 \times -3 \times -4}{3 \times 2 \times 1} = \frac{-24}{6} = -4$$

$$\frac{(-2)(-3)(-4)(-5)}{4!} = \frac{-2 \times -3 \times -4 \times -5}{4 \times 3 \times 2 \times 1} = \frac{120}{24} = 5$$

So the expansion simplifies to
$1 - 2x + 3x^2 - 4x^3 + 5x^4 - \ldots$

(b) $\ldots - 6x^5 + 7x^6 - 8x^7$

(c) If n is negative then the factors in the product
$n(n-1)(n-2)(n-3)\ldots$ are all non-zero, no
matter how many factors there are. Hence a
non-zero coefficient of x^k exists for all values
of k. The powers of x form the infinite
sequence x^0, x^1, x^2, x^3, x^4, $\ldots$ so the expansion
can go on for ever. However, if we try to
continue the expansion of $(1 + ax)^n$ when n is
a positive integer, we will find that after the
last term which is $a^n x^n$ the coefficients will all
be zero (as $(n - n)$ which is 0 will be a factor
of the product $n(n-1)(n-2)(n-3)\ldots$).
So the number of terms is finite.

(d) $15x^{14}$

B2 (a) (i) $0.694\,444$

(ii)
Expansion	Value
1	1
$1 - 2x$	0.6
$1 - 2x + 3x^2$	0.72
$1 - 2x + 3x^2 - 4x^3$	0.688
$1 - 2x + 3x^2 - 4x^3 + 5x^4$	0.696
$1 - 2x + 3x^2 - 4x^3 + 5x^4 - 6x^5$	0.694\,08
$1 - 2x + 3x^2 - 4x^3 + 5x^4 - 6x^5 + 7x^6$	0.694\,528
$1 - 2x + 3x^2 - 4x^3 + 5x^4 - 6x^5 + 7x^6 - 8x^7$	0.694\,4256

(iii) The values appear to be converging
towards a limit which looks as though it is
the value of $(1 + x)^{-2}$ when $x = 0.2$,
i.e. $0.694\,444\ldots$

(b) (i) 1

(ii) Values when $x = -2$ are:
$1, 5, 17, 49, 129, 321, 769, 1793$
The values are diverging and not
converging towards the value of $(1 + x)^{-2}$
when $x = -2$, i.e. 1.

(c) (i) $0.277\,008$

(ii) Values (rounded to 6 d.p. where
appropriate) when $x = 0.9$ are:
$1, -0.8, 1.63, -1.286, 1.9945, -1.548\,44$,
$2.171\,647, -1.654\,728$
The values do not appear to be converging
but it is not possible to deduce from the
expansion that values will continue to
diverge.

B3 (a) The expansion does in fact converge to the
value of $(1 + x)^{-2}$ when $x = 0.9$, i.e. $0.277\,008$
(to 6 d.p.), though you need to evaluate the
sum of many terms in the expansion before
you can conjecture this with confidence.

(b) When $-1 < x < 1$, then the expansion will
converge. For values close to 0, the
convergence is much faster than for those
values near -1 or 1.

B4

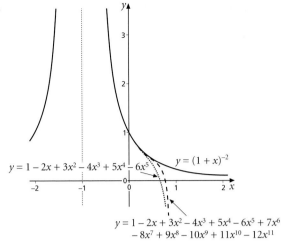

$y = 1 - 2x + 3x^2 - 4x^3 + 5x^4 - 6x^5$

$y = (1 + x)^{-2}$

$y = 1 - 2x + 3x^2 - 4x^3 + 5x^4 - 6x^5 + 7x^6$
$- 8x^7 + 9x^8 - 10x^9 + 11x^{10} - 12x^{11}$

Comments such as:
The graphs of $y = f(x)$ and $y = g(x)$ are very close between about -0.5 and 0.5. The graph of $y = f(x)$ is very close to the graph of $y = h(x)$ over a larger interval between about -0.7 and 0.7.

B5 (a) Using the binomial theorem gives

$$(1 - 2x)^{-1} = 1 + (-1)(-2x) + \frac{(-1)(-2)}{2!}(-2x)^2$$

$$+ \frac{(-1)(-2)(-3)}{3!}(-2x)^3 + \frac{(-1)(-2)(-3)(-4)}{4!}(-2x)^4$$

$$+ \frac{(-1)(-2)(-3)(-4)(-5)}{5!}(-2x)^5 + \dots$$

$$= 1 + 2x + 4x^2 + 8x^3 + 16x^4 + 32x^5 + \dots$$

(b) $2048x^{11}$

(c) (i) -2.5

(ii) Values when $x = 0.7$ are:
1, 2.4, 4.36, 7.104, 10.9456, … and further rows in the spreadsheet show that the values are diverging and not converging towards the value of $(1 - 2x)^{-1}$ when $x = 0.7$, i.e. -2.5.

(d) When $-0.5 < x < 0.5$, then the expansion will converge to the value of $(1 - 2x)^{-1}$.

B6 (a) The coefficients for x^2 and x^3 are

$$\frac{\left(\frac{1}{2}\right)\left(-\frac{1}{2}\right)}{2!} = \frac{\frac{1}{2} \times -\frac{1}{2}}{2 \times 1} = \frac{-\frac{1}{4}}{2} = -\frac{1}{8}$$

$$\frac{\left(\frac{1}{2}\right)\left(-\frac{1}{2}\right)\left(-\frac{3}{2}\right)}{3!} = \frac{\frac{1}{2} \times -\frac{1}{2} \times -\frac{3}{2}}{3 \times 2 \times 1} = \frac{\frac{3}{8}}{6} = \frac{3}{48} = \frac{1}{16}$$

So the expansion simplifies to
$1 + \frac{1}{2}x - \frac{1}{8}x^2 + \frac{1}{16}x^3 + \dots$

(b) $\dots - \frac{5}{128}x^4 + \frac{7}{256}x^5$

(c) (i) $1.341\,641$ (to 6 d.p.)   **(ii)** 1.352

(d)

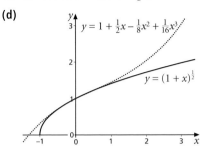

$y = 1 + \frac{1}{2}x - \frac{1}{8}x^2 + \frac{1}{16}x^3$

$y = (1 + x)^{\frac{1}{2}}$

As shown above, between $x = -1$ and $x = 1$ the graphs of $y = (1 + x)^{\frac{1}{2}}$ and $y = 1 + \frac{1}{2}x - \frac{1}{8}x^2 + \frac{1}{16}x^3$ are close.

B7 (a) $|2x| < 1 \;\Rightarrow\; -1 < 2x < 1$
$\Rightarrow\; -\frac{1}{2} < x < \frac{1}{2}$
$\Rightarrow\; |x| < \frac{1}{2}$

(b) $|-4x| < 1 \;\Rightarrow\; -1 < -4x < 1$
$\Rightarrow\; 1 > 4x > -1$
$\Rightarrow\; -1 < 4x < 1$
$\Rightarrow\; -\frac{1}{4} < x < \frac{1}{4}$
$\Rightarrow\; |x| < \frac{1}{4}$

(c) $\left|\frac{1}{3}x\right| < 1 \;\Rightarrow\; -1 < \frac{1}{3}x < 1$
$\Rightarrow\; -3 < x < 3$
$\Rightarrow\; |x| < 3$

(d) $\left|\frac{3}{4}x\right| < 1 \;\Rightarrow\; -1 < \frac{3}{4}x < 1$
$\Rightarrow\; -\frac{4}{3} < x < \frac{4}{3}$
$\Rightarrow\; |x| < \frac{4}{3}$

Exercise B (p 41)

1 (a) $1 - x + x^2 - x^3$ $|x| < 1$

(b) $1 + 2x + 3x^2 + 4x^3$ $|x| < 1$

(c) $1 - 8x + 40x^2 - 160x^3$ $|x| < \frac{1}{2}$

(d) $1 - x + \frac{3}{4}x^2 - \frac{1}{2}x^3$ $|x| < 2$

(e) $1 - 3x + 9x^2 - 27x^3$ $|x| < \frac{1}{3}$

(f) $1 + \frac{2}{3}x + \frac{1}{3}x^2 + \frac{4}{27}x^3$ $|x| < 3$

(g) $1 - \frac{1}{2}x - \frac{1}{8}x^2 - \frac{1}{16}x^3$ $|x| < 1$

(h) $1 + \frac{1}{8}x - \frac{1}{128}x^2 + \frac{1}{1024}x^3$ $|x| < 4$

(i) $1 - x - \frac{3}{2}x^2 - \frac{5}{2}x^3$ $|x| < \frac{1}{2}$

2 (a) Using the binomial theorem,

$$(1+3x)^{\frac{1}{3}} = 1 + \left(\tfrac{1}{3}\right)(3x) + \frac{\left(\tfrac{1}{3}\right)\left(-\tfrac{2}{3}\right)}{2!}(3x)^2$$

$$+ \frac{\left(\tfrac{1}{3}\right)\left(-\tfrac{2}{3}\right)\left(-\tfrac{5}{3}\right)}{3!}(3x)^3 + \ldots = 1 + x - x^2 + \tfrac{5}{3}x^3 - \ldots$$

(b) $|x| < \tfrac{1}{3}$

3 (a) $1 + 2x - 2x^2 + 4x^3 - 10x^4 + 28x^5 - 84x^6$

(b) The expansion is valid when $|x| < \tfrac{1}{4}$.
$\sqrt{1.4} = (1 + 4 \times 0.1)^{\frac{1}{2}}$ which is $(1 + 4x)^{\frac{1}{2}}$ when
$x = 0.1$. As $|0.1| < \tfrac{1}{4}$, the expansion will be
valid for this value.
So $\sqrt{1.4} \approx 1 + 2(0.1) - 2(0.1)^2 + 4(0.1)^3$
$- 10(0.1)^4 + 28(0.1)^5 - 84(0.1)^6 = 1 + 0.2$
$- 0.02 + 0.004 - 0.001 + 0.000\,28 - 0.000\,084$.
Now the first six terms give $1.183\,28$ and
subsequent terms will not alter the value of
the first two decimal places. Hence
$\sqrt{1.4} = 1.18$ (to 2 d.p.)

4 (a) $1 + 3x + \tfrac{3}{2}x^2$

(b) $1.8^{\frac{3}{2}} = (1 + 2 \times 0.4)^{\frac{3}{2}} \approx 1 + 3(0.4) + \tfrac{3}{2}(0.4)^2$
$= 2.44$
Using a calculator gives $1.8^{\frac{3}{2}} = 2.414\,953\,4\ldots$,
which agrees with the approximate value
when both values are rounded to 2 s.f. (to give
2.4). So the approximate value is accurate to
2 s.f.

5 (a) $1 + \tfrac{1}{4}x - \tfrac{1}{32}x^2$

(b) (i) $4 + x - \tfrac{1}{8}x^2$ **(ii)** $|x| < 2$

6 $\tfrac{1}{648}$

7 $(8 - 3x)^{-\frac{1}{3}} = 8^{-\frac{1}{3}}(1 - \tfrac{3}{8}x)^{-\frac{1}{3}} = \tfrac{1}{2}(1 - \tfrac{3}{8}x)^{-\frac{1}{3}}$
so the coefficient of x^3 is
$$\frac{\tfrac{1}{2}\left(-\tfrac{1}{3}\right)\left(-\tfrac{4}{3}\right)\left(-\tfrac{7}{3}\right)\left(-\tfrac{3}{8}\right)^3}{3!} = \frac{7}{1536}$$

8 (a) $\tfrac{1}{2} - \tfrac{1}{4}x + \tfrac{1}{8}x^2$ $|x| < 2$

(b) $\tfrac{1}{64} + \tfrac{3}{256}x + \tfrac{3}{512}x^2$ $|x| < 4$

(c) $\tfrac{1}{36} - \tfrac{1}{36}x + \tfrac{1}{48}x^2$ $|x| < 2$

(d) $2 + \tfrac{1}{4}x - \tfrac{1}{64}x^2$ $|x| < 4$

(e) $\tfrac{1}{3} - \tfrac{2}{9}x + \tfrac{4}{27}x^2$ $|x| < \tfrac{3}{2}$

(f) $2 - \tfrac{4}{5}x - \tfrac{16}{25}x^2$ $|x| < \tfrac{1}{2}$

C Multiplying to obtain expansions

Exercise C (p 42)

1 (a) $5 + 10x + 15x^2 + 20x^3$

(b) $x - 3x^2 + 9x^3$

(c) $1 + 3x + 8x^2 + 26x^3$

(d) $1 + 7x + 36x^2 + 162x^3$

2 (a) (i) $(1 - 3x)^{-1} = 1 + (-1)(-3x) + \dfrac{(-1)(-2)}{2!}(-3x)^2 + \ldots$

     $\approx 1 + 3x + 9x^2$

(ii) $|x| < \tfrac{1}{3}$

(b) (i) $(1 + x)^{-4} = 1 + (-4)x + \dfrac{(-4)(-5)}{2!}x^2 + \ldots$

     $\approx 1 - 4x + 10x^2$

(ii) $|x| < 1$

(c) (i) $(1 + 3x + 9x^2 + \ldots)(1 - 4x + 10x^2 - \ldots)$
$= 1 - 4x + 10x^2 + \ldots + 3x - 12x^2 + \ldots$
     $+ 9x^2 + \ldots$
$= 1 - x + 7x^2 + \ldots$
$\approx 1 - x + 7x^2$

(ii) The expansion is valid for values of x that
satisfy both $|x| < \tfrac{1}{3}$ and $|x| < 1$, that is,
which satisfy $|x| < \tfrac{1}{3}$.

D Adding (using partial fractions) to obtain expansions

Exercise D (p 44)

1 (a) $\dfrac{3}{1-x} + \dfrac{1}{1+2x}$

(b) $4 + x + 7x^2$

(c) $\dfrac{1}{1-x}$ is valid for $|x| < 1$ and $\dfrac{1}{1+2x}$ is valid for
$|x| < \tfrac{1}{2}$ so the expansion for f(x) is valid for
values of x that satisfy both inequalities, i.e.
for $|x| < \tfrac{1}{2}$.

2 (a) $\dfrac{2}{2+x} + \dfrac{3}{1-2x}$

(b) (i) $\dfrac{1}{2+x} = (2+x)^{-1} = 2^{-1}\left(1 + \tfrac{1}{2}x\right)^{-1}$

$= \tfrac{1}{2}\left(1 + \tfrac{1}{2}x\right)^{-1}$

$= \tfrac{1}{2}\left(1 + (-1)\left(\tfrac{1}{2}x\right) + \dfrac{(-1)(-2)}{2!}\left(\tfrac{1}{2}x\right)^2 + \ldots\right)$

$= \tfrac{1}{2}\left(1 - \tfrac{1}{2}x + \tfrac{1}{4}x^2 - \ldots\right)$

$= \tfrac{1}{2} - \tfrac{1}{4}x + \tfrac{1}{8}x^2 - \ldots$

so the first three terms are $\tfrac{1}{2} - \tfrac{1}{4}x + \tfrac{1}{8}x^2$.

(ii) $1 + 2x + 4x^2$

(c) $4 + \tfrac{11}{2}x + \tfrac{49}{4}x^2$

(d) $|x| < \tfrac{1}{2}$

3 (a) $2 - x + 25x^2 - 37x^3$ $\qquad |x| < \tfrac{1}{4}$

(b) $\tfrac{7}{3} - \tfrac{53}{9}x + \tfrac{487}{27}x^2 - \tfrac{4373}{81}x^3$ $\quad |x| < \tfrac{1}{3}$

(c) $3 - \tfrac{4}{3}x + \tfrac{22}{9}x^2 - \tfrac{46}{27}x^3$ $\qquad |x| < 1$

4 $\dfrac{11x - 3}{(4-3x)(1+x)} \equiv \dfrac{A}{4-3x} + \dfrac{B}{1+x}$

$\Rightarrow A(1+x) + B(4-3x) \equiv 11x - 3$

$x = -1 \Rightarrow 7B = -14 \Rightarrow B = -2$

$x = \tfrac{4}{3} \Rightarrow \tfrac{7}{3}A = \tfrac{35}{3} \Rightarrow A = 5$

So $\dfrac{11x - 3}{(4-3x)(1+x)} \equiv \dfrac{5}{4-3x} - \dfrac{2}{1+x}$

$\dfrac{5}{4-3x} = 5(4-3x)^{-1} = \tfrac{5}{4}\left(1 - \tfrac{3}{4}x\right)^{-1}$

$= \tfrac{5}{4}\left(1 + (-1)\left(-\tfrac{3}{4}x\right) + \dfrac{(-1)(-2)}{2!}\left(-\tfrac{3}{4}x\right)^2 + \ldots\right)$

$= \tfrac{5}{4}\left(1 + \tfrac{3}{4}x + \tfrac{9}{16}x^2 + \ldots\right)$

$= \tfrac{5}{4} + \tfrac{15}{16}x + \tfrac{45}{64}x^2 + \ldots$

$\dfrac{2}{1+x} = 2(1+x)^{-1} = 2\left(1 - x + x^2 - \ldots\right)$

$= 2 - 2x + 2x^2 - \ldots$

so $\dfrac{5}{4-3x} - \dfrac{2}{1+x}$

$= \left(\tfrac{5}{4} + \tfrac{15}{16}x + \tfrac{45}{64}x^2 + \ldots\right) - \left(2 - 2x + 2x^2 - \ldots\right)$

$= -\tfrac{3}{4} + \tfrac{47}{16}x - \tfrac{83}{64}x^2 + \ldots$

So the first three terms of the expansion are

$-\tfrac{3}{4} + \tfrac{47}{16}x - \tfrac{83}{64}x^2$ as required.

Mixed questions (p 44)

1 (a) $4 - 24x + 96x^2$ $\qquad$ **(b)** $|x| < \tfrac{1}{2}$

2 (a) $10 - 20x - 20x^2 - 40x^3$

(b) (i) $x = 0.1$

(ii) $\sqrt{60} = 60^{\frac{1}{2}} = (100 - 400 \times 0.1)^{\frac{1}{2}}$

$\approx 10 - 20(0.1) - 20(0.1)^2 - 40(0.1)^3$

$= 7.76$

(iii) From a calculator, $\sqrt{60} = 7.745\,966\,69\ldots$, which agrees with the approximate value when both values are rounded to 1 s.f. (to give 8). So the approximate value is accurate to 1 s.f.

3 $a = 5,\ n = -2$

4 $(16 + 192x)^{\frac{3}{4}} = 16^{\frac{3}{4}}(1 + 12x)^{\frac{3}{4}}$

$= 8(1 + 12x)^{\frac{3}{4}}$

so the coefficient of x^3 is $\dfrac{8\left(\tfrac{3}{4}\right)\left(-\tfrac{1}{4}\right)\left(-\tfrac{5}{4}\right)12^3}{3!} = 540$

5 (a) $1 - x + x^2 - x^3$

(b) $(1 + x)^{-1} = \dfrac{1}{1+x}$; the integral of this is $\ln(1 + x)$. The integral of the first four terms of the expansion is $x - \tfrac{1}{2}x^2 + \tfrac{1}{3}x^3 - \tfrac{1}{4}x^4$ so the series expansion for $\ln(1 + x)$ as far as the term in x^4 is $x - \tfrac{1}{2}x^2 + \tfrac{1}{3}x^3 - \tfrac{1}{4}x^4$.

(c) No, $\ln(1 + x) = \ln 3$ when $x = 2$ and the expansion is only valid for $|x| < 1$.

6 (a) $1 + 2x + 4x^2 + 8x^3$

(b) $\dfrac{1-x}{1-2x} = (1 - x)(1 - 2x)^{-1}$

$= (1 - x)(1 + 2x + 4x^2 + 8x^3 + \ldots)$

$= 1 + 2x + 4x^2 + 8x^3 + \ldots - x - 2x^2 - 4x^3 - \ldots$

$= 1 + x + 2x^2 + 4x^3 + \ldots$

So the first four terms in the expansion are $1 + x + 2x^2 + 4x^3$.

(c) $\dfrac{99}{98} = \dfrac{0.99}{0.98} = \dfrac{1 - 0.01}{1 - 0.02} = \dfrac{1 - x}{1 - 2x}$ when $x = 0.01$.

So $\dfrac{99}{98} \approx 1 + (0.01) + 2(0.01)^2 + 4(0.01)^3$

$= 1 + 0.01 + 0.0002 + 0.000004$

$= 1.010\,20$ (to 5 d.p.)

(d) $0.010\,20$

Test yourself (p 45)

1 (a) $p = 24, q = 80$

  (b) (i) $1 + 3x + 6x^2$ **(ii)** $|x| < \frac{1}{2}$

2 (a) $A = 1, B = 27$

  (b) $10 - 2x + 2x^2 + \frac{2}{3}x^3$

3 (a) When $x = \frac{1}{15}$, $(1 + 5x)^{-\frac{1}{2}} = (1 + \frac{5}{15})^{-\frac{1}{2}}$

$$= \left(\tfrac{4}{3}\right)^{-\frac{1}{2}} = \left(\frac{2}{\sqrt{3}}\right)^{-1} = \frac{\sqrt{3}}{2}$$

  which is the exact value of $\sin 60°$

  (b) $1 - \frac{5}{2}x + \frac{75}{8}x^2 - \frac{625}{16}x^3$

  (c) Using the answer in (b), when $x = \frac{1}{15}$

$$\sin 60° \approx 1 - \frac{5}{2} \times \frac{1}{15} + \frac{75}{8} \times \frac{1}{15^2} - \frac{625}{16} \times \frac{1}{15^3}$$

$$= \frac{373}{432} \approx 0.8634$$

  (d) $0.002\,599$ (to 4 s.f.)

4 (a) $1 - 2x - 8x^2 - 48x^3$

  (b) $(1 - 10 \times 0.001)^{\frac{1}{5}} = 0.99^{\frac{1}{5}}$

$$\approx 1 - (2 \times 0.001) - (8 \times 0.001^2) - (48 \times 0.001^3)$$

$$= 0.997\,991\,952$$

  So $\sqrt[5]{0.99} = 0.997\,99$ to 5 s.f. (further terms in the expansion will not affect the value of the fifth significant figure).

  Hence $\sqrt[5]{99\,000} = \sqrt[5]{0.99 \times 100\,000}$

$$= \sqrt[5]{0.99} \times \sqrt[5]{100\,000}$$

$$= 10 \times \sqrt[5]{0.99}$$

$$= 10 \times 0.997\,99 \text{ (to 5 s.f.)}$$

$$= 9.9799 \text{ (to 5 s.f.)}$$

5 $\frac{5}{32}$

4 Differentiation

A Functions defined parametrically (p 46)

A1 $\frac{3}{2}t$

A2 (a) $t + \frac{1}{2}$ **(b)** $-\dfrac{1}{3t^2}$ **(c)** $-\cot t$

A3 $y - y_1 = m(x - x_1)$, $(x_1, y_1) = (6, 5)$, $m = \frac{4}{3}$

  So $y - 5 = \frac{4}{3}(x - 6)$

  $\Rightarrow y - 5 = \frac{4}{3}x - 8$

  $\Rightarrow \quad y = \frac{4}{3}x - 3$

A4 $y - 5 = -\frac{3}{4}(x - 6)$

  $\Rightarrow y - 5 = -\frac{3}{4}x + \frac{9}{2}$

  $\Rightarrow \quad y = -\frac{3}{4}x + \frac{19}{2}$

A5 (a) $-\dfrac{1}{2t^2}$

  (b) $\left(8, \frac{1}{4}\right)$

  (c) Gradient $= m = -\dfrac{1}{2 \times 4^2} = -\dfrac{1}{32}$

  Using $y - y_1 = m(x - x_1)$

$$y - \tfrac{1}{4} = -\tfrac{1}{32}(x - 8)$$

  $\Rightarrow \quad y - \frac{1}{4} = -\frac{1}{32}x + \frac{1}{4}$

  $\Rightarrow \quad\quad y = -\frac{1}{32}x + \frac{1}{2}$

  (d) $x + 32y = 16$

A6 (a) $\dfrac{dx}{dt} = \cos t, \ \dfrac{dy}{dt} = \sin t$

$$\frac{dy}{dx} = \frac{\dfrac{dy}{dt}}{\dfrac{dx}{dt}} = \frac{\sin t}{\cos t} = \tan t$$

  (b) $x + y = 1$

Exercise A (p 49)

1 (a) $\dfrac{dy}{dt} = 3t^2, \ \dfrac{dx}{dt} = 2t$

$$\frac{dy}{dx} = \frac{\dfrac{dy}{dt}}{\dfrac{dx}{dt}} = \frac{3t^2}{2t} = \frac{3}{2}t$$

  (b) (i) $(16, 64)$ **(ii)** $y = 6x - 32$

  (iii) $y = -\frac{1}{6}x + \frac{200}{3}$

2 (a) $-\frac{1}{4}t^{\frac{3}{2}}$

(b) When $t = 1$, $x = 3$, $y = 2$, $\frac{dy}{dx} = -\frac{1}{4}$

$$y - 2 = -\frac{1}{4}(x - 3)$$
$$\Rightarrow \quad y - 2 = -\frac{1}{4}x + \frac{3}{4}$$
$$\Rightarrow \quad y = -\frac{1}{4}x + \frac{11}{4}$$

(c) $y = 4x - 10$

3 (a) $\frac{dy}{d\theta} = 2\cos\theta$, $\frac{dx}{d\theta} = -\sin\theta$

$$\frac{dy}{dx} = \frac{\frac{dy}{d\theta}}{\frac{dx}{d\theta}} = \frac{2\cos\theta}{-\sin\theta} = -\frac{2}{\tan\theta}$$

(b) $\left(\frac{1}{\sqrt{2}}, \sqrt{2}\right)$

(c) $m = -\frac{2}{\tan\frac{\pi}{4}} = -2$

$$y - \sqrt{2} = -2\left(x - \frac{1}{\sqrt{2}}\right)$$
$$\Rightarrow y - \sqrt{2} = -2x + \sqrt{2}$$
$$\Rightarrow 2x + y = 2\sqrt{2}$$

4 The coordinates of P are $\left(3\cos\left(\frac{1}{3}\pi\right), 4\sin\left(\frac{1}{3}\pi\right) - 1\right)$, namely $\left(\frac{3}{2}, 2\sqrt{3} - 1\right)$

$\frac{dx}{dt} = -3\sin t$, $\frac{dy}{dt} = 4\cos t$ giving $\frac{dy}{dx} = -\frac{4\cos t}{3\sin t}$

So at $t = \frac{1}{3}\pi$ the gradient of the curve is

$$-\frac{4\cos\left(\frac{1}{3}\pi\right)}{3\sin\left(\frac{1}{3}\pi\right)} = -\frac{4 \times \frac{1}{2}}{3 \times \frac{\sqrt{3}}{2}} = -\frac{4}{3\sqrt{3}}.$$

So the gradient of the normal is $\frac{3\sqrt{3}}{4}$.

So the equation of the normal is

$$y - \left(2\sqrt{3} - 1\right) = \frac{3\sqrt{3}}{4}\left(x - \frac{3}{2}\right)$$
$$\Rightarrow \quad y - 2\sqrt{3} + 1 = \frac{3\sqrt{3}}{4}x - \frac{9\sqrt{3}}{8}$$
$$\Rightarrow 8y - 16\sqrt{3} + 8 = 6\sqrt{3}x - 9\sqrt{3}$$
$$\Rightarrow \qquad\qquad 8y = 6\sqrt{3}x + 7\sqrt{3} - 8$$

as required.

5 (a) $\dfrac{2t}{1 + \frac{1}{t^2}} = \dfrac{2t^3}{t^2 + 1}$

(b) (i) $4x - 20y + 11 = 0$

(ii) $20x + 4y + 29 = 0$

6 (a) $\frac{dx}{d\theta} = 4\cos\theta(-\sin\theta) = -2(2\sin\theta\cos\theta) = -2\sin$

$\frac{dy}{d\theta} = 2\cos 2\theta$ giving $\frac{dy}{dx} = \frac{2\cos 2\theta}{-2\sin 2\theta} = -\frac{\cos}{\sin}$

$= -\cot 2\theta$ as required

(b) The coordinates of the point where $\theta = \frac{\pi}{6}$ are

$$\left(2\left(\cos\frac{\pi}{6}\right)^2, \sin\left(2 \times \frac{\pi}{6}\right)\right) = \left(2\left(\frac{\sqrt{3}}{2}\right)^2, \frac{\sqrt{3}}{2}\right)$$
$$= \left(\frac{3}{2}, \frac{\sqrt{3}}{2}\right)$$

The gradient of the tangent at this point is

$$-\cot\left(2 \times \frac{\pi}{6}\right) = -\frac{1}{\tan\frac{\pi}{3}} = -\frac{1}{\sqrt{3}}.$$

So the equation of the tangent is

$$y - \frac{\sqrt{3}}{2} = -\frac{1}{\sqrt{3}}\left(x - \frac{3}{2}\right) \Rightarrow \sqrt{3}y - \frac{3}{2} = -x + \frac{3}{2}$$
$$\Rightarrow \sqrt{3}y + x = 3 \text{ as required.}$$

7 $x = 1$

B Functions defined implicitly (p 50)

B1 (a) $\frac{d}{dx}(\ln x) = \frac{1}{x}$

(b) $\frac{d}{dx}(x + \sin 3x) = 1 + 3\cos 3x$

B2 (a) $2\frac{dy}{dx} + 3 + \frac{dy}{dx} = 0$

$3\frac{dy}{dx} + 3 = 0$ so $\frac{dy}{dx} = -1$

(b) $x\frac{dy}{dx} + y + \frac{dy}{dx} = 0$

$$\Rightarrow (x + 1)\frac{dy}{dx} + y = 0$$
$$\Rightarrow \qquad (x + 1)\frac{dy}{dx} = -y$$
$$\Rightarrow \qquad \frac{dy}{dx} = -\frac{y}{x + 1}$$

B3 (a) $3x^2 + \frac{dy}{dx}$ **(b)** $4x^3 - \frac{dy}{dx}$ **(c)** $-\frac{1}{x^2} + 2$

B4 (a) $x^3\frac{dy}{dx} + 3x^2y$ **(b)** $x^4\frac{dy}{dx} + 4x^3y$ **(c)** $e^x\frac{dy}{dx} + e$

B5 (a) Differentiate with respect to x:

$$x^2\frac{dy}{dx} + 2xy + 1 + \frac{dy}{dx} = 0$$
$$\Rightarrow (x^2 + 1)\frac{dy}{dx} + 2xy + 1 = 0$$

(b) $x^2y + x + y = (1)^2 \times 4 + 1 + 4 = 4 + 1 + 4 = 9$

Gradient: $-\frac{9}{2}$

B6 (a) $3y^2\dfrac{dy}{dx}$ (b) $4y^3\dfrac{dy}{dx}$ (c) $3(y-2)^2\dfrac{dy}{dx}$

(d) $4(2y+1)\dfrac{dy}{dx}$ (e) $-2y^{-3}\dfrac{dy}{dx}$

B7 (a) $3xy^2\dfrac{dy}{dx} + y^3$ (b) $2x^2y\dfrac{dy}{dx} + 2xy^2$

(c) $4x^3y^3\dfrac{dy}{dx} + 3x^2y^4$ (d) $2x(y-1)\dfrac{dy}{dx} + (y-1)^2$

B8 (a) Differentiating with respect to x,

$2y\dfrac{dy}{dx} + 1 + \dfrac{dy}{dx} = 0 \Rightarrow (2y+1)\dfrac{dy}{dx} + 1 = 0$

(b) $2^2 + 1 + 2 = 7$ so $(1, 2)$ lies on the curve.

At $(1, 2)$, $\dfrac{dy}{dx} = -\frac{1}{5}$

B9 (a) $\dfrac{d}{dx}(xy^2) + \dfrac{dy}{dx} = 0$

$x\dfrac{d}{dx}(y^2) + y^2\dfrac{d(x)}{dx} + \dfrac{dy}{dx} = 0$

$x(2y)\dfrac{dy}{dx} + y^2 + \dfrac{dy}{dx} = 0$

$(2xy + 1)\dfrac{dy}{dx} + y^2 = 0$

$\dfrac{dy}{dx} = -\dfrac{y^2}{2xy + 1}$

(b) $1 \times (2)^2 + 2 = 6$ so $(1, 2)$ lies on the curve.

At $(1, 2)$, $\dfrac{dy}{dx} = -\frac{4}{5}$

Exercise B (p 53)

1 (a) $\dfrac{d}{dx}(x^2 + xy + y) = 0$

$2x + \left(x\dfrac{dy}{dx} + y\right) + \dfrac{dy}{dx} = 0$

$2x + y + (x + 1)\dfrac{dy}{dx} = 0$

$\dfrac{dy}{dx} = -\dfrac{2x + y}{x + 1}$

(b) Tangent: $3y + 5x = 13$
Normal: $5y - 3x = -1$

2 (a) (i) $2x - y - 1 = 0$

(ii) $x + 2y - 3 = 0$

(b) (i) $13x + 3y - 29 = 0$

(ii) $3x - 13y + 7 = 0$

(c) (i) $5x - 3y - 13 = 0$

(ii) $3x + 5y - 35 = 0$

(d) (i) $7x + 15y - 36 = 0$

(ii) $15x - 7y - 38 = 0$

(e) (i) $6x + y - 13 = 0$

(ii) $x - 6y + 4 = 0$

(f) (i) $x + 3y - 13 = 0$

(ii) $3x - y - 9 = 0$

3 (a) $(1, 2)$, $(1, 5)$ (b) $\frac{4}{3}$, $-\frac{7}{3}$

4 (a) $(3, 1)$, $(-4, 1)$ (b) $-\frac{7}{5}$, $-\frac{7}{2}$

(c) $7x + 5y - 26 = 0$, $7x + 2y + 26 = 0$

Mixed questions (p 54)

1 (a) (i) $-\dfrac{1}{2t^2}$ (ii) $-\frac{1}{8}$

(b) $\left(5, \frac{3}{2}\right)$

(c) (i) $2t = x - 1$ so $t = \frac{1}{2}(x - 1)$

$y = 1 + \dfrac{1}{\frac{1}{2}(x-1)}$

$\Rightarrow \frac{1}{2}(x-1)y = \frac{1}{2}(x-1) + 1$

$\Rightarrow (x-1)y = (x-1) + 2$

$\Rightarrow xy - y = x + 1$ so $xy - x - y = 1$

(ii) $-\frac{1}{8}$

(d) (a) (ii) and (c) (ii) have the same answer.

2 (a)

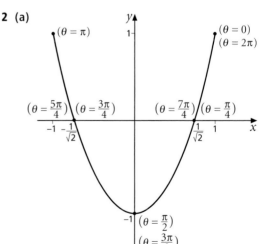

(b) The curve repeats itself because
$\cos(\theta + 2\pi) = \cos\theta$ and $\cos(2\theta + 2\pi) = \cos 2\theta$.

(c) $\cos 2\theta = 2\cos^2\theta - 1$, so $y = 2x^2 - 1$

The graph is only part of $y = 2x^2 - 1$ because
$x = \cos\theta$, so $-1 \le x \le 1$.

(d) $\dfrac{2\sin 2\theta}{\sin \theta} = 4\cos\theta$

(e) $2\sqrt{2}x - y - 2 = 0$

(f) $\sqrt{2}x + 4y - 1 = 0$

3 (a) $-\dfrac{x-2}{y-3} = \dfrac{2-x}{y-3}$

(b) $(2, -2)\ (6, 6)$

(c) $y = -2,\quad 3y + 4x = 42$

(d) $x = 2,\quad 4y - 3x = 6$

(e) The centre of the circle is $(2, 3)$
$(2, 3)$ lies on the line $x = 2$
$4(3) - 3(2) = 6$ so $(2, 3)$ lies on the line
$4y - 3x = 6$

4 (a) $x^{\frac{1}{2}} + y^{\frac{1}{2}} = 5$

$\frac{1}{2}x^{-\frac{1}{2}} + \frac{1}{2}y^{-\frac{1}{2}}\dfrac{dy}{dx} = 0$

$\Rightarrow \dfrac{dy}{dx} = -\dfrac{x^{-\frac{1}{2}}}{y^{-\frac{1}{2}}} = -\dfrac{y^{\frac{1}{2}}}{x^{\frac{1}{2}}} = -\sqrt{\dfrac{y}{x}}$

(b) $2x + 3y = 30$

5 (a) $\dfrac{\sqrt{3}}{2}$ **(b)** $2\sqrt{3}y + 4x = 3\sqrt{3} + 4$

(c) $x = 2$

6 (a) Differentiating with respect to x,

$(x^2 - 1)\dfrac{dy}{dx} + 2xy = 2x$

$\Rightarrow (x^2 - 1)\dfrac{dy}{dx} = 2x - 2xy = 2x(1 - y)$

(b) When $x = 2$, $y = \frac{5}{3}$ and the gradient is $-\frac{8}{9}$.

(b) $(-2, 1)$ and $(2, -1)$

4 (a) $\dfrac{dx}{dt} = 12\sin^2 t\cos t,\ \dfrac{dy}{dt} = -2\sin 2t$ giving

$\dfrac{dx}{dy} = \dfrac{12\sin^2 t\cos t}{-2\sin 2t} = \dfrac{12\sin^2 t\cos t}{-4\sin t\cos t} = -3\sin t$

as required

(b) $y = \frac{3}{2}x - \frac{1}{4}$ or $4y = 6x - 1$ or equivalent

Test yourself (p 55)

1 Tangent: $7x + 10y - 24 = 0$
Normal: $10x - 7y - 13 = 0$

2 $-\sqrt{2}$

3 (a) Differentiating gives
$14x + 48y + 48x\dfrac{dy}{dx} - 14y\dfrac{dy}{dx} = 0$
which simplifies to
$7x + 24y + \dfrac{dy}{dx}(24x - 7y) = 0$

$\dfrac{dy}{dx} = \frac{2}{11}$ gives $7x + 24y + \frac{2}{11}(24x - 7y) = 0$

$\Rightarrow 77x + 264y + 48x - 14y = 0$

$\Rightarrow 125x + 250y = 0$

$\Rightarrow x + 2y = 0$ as required

5 Integration 1

A Review

Exercise A (p 56)

1 (a) $\frac{1}{3}x^3 - \frac{1}{2}x^2 + c$ (b) $x^5 + x + c$

(c) $\frac{1}{2}x^4 - \frac{1}{6}x^3 - 3x + c$

2 $-x^{-3} + c$ or $-\dfrac{1}{x^3} + c$

3 (a) $-x^{-1} - x^{-2} + c$ or $-\dfrac{1}{x} - \dfrac{1}{x^2} + c$

(b) $x + \frac{1}{3}x^{-1} + c$ or $x + \dfrac{1}{3x} + c$

(c) $-\frac{3}{5}x^{-2} + c$ or $-\dfrac{3}{5x^2} + c$

4 $\displaystyle\int_1^2 \left(\frac{3}{5}x^2 - \frac{1}{x^2}\right) dx$

$= \displaystyle\int_1^2 \left(\frac{3}{5}x^2 - x^{-2}\right) dx$

$= \left[\dfrac{3}{5} \times \dfrac{x^3}{3} - \dfrac{x^{-1}}{-1}\right]_1^2$

$= \left[\dfrac{1}{5}x^3 + \dfrac{1}{x}\right]_1^2$

$= \left(\frac{8}{5} + \frac{1}{2}\right) - \left(\frac{1}{5} + 1\right)$

$= \frac{9}{10}$

5 (a) $2x^{\frac{3}{2}} + c$ or $2\sqrt{x^3} + c$

(b) $\frac{3}{7}x^{\frac{7}{3}} + c$ or $\frac{3}{7}\sqrt[3]{x^7} + c$

(c) $\frac{2}{3}x^{\frac{3}{2}} + c$ or $\frac{2}{3}\sqrt{x^3} + c$

(d) $x^{\frac{1}{2}} + c$ or $\sqrt{x} + c$

(e) $\frac{2}{5}x^{\frac{5}{2}} + c$ or $\frac{2}{5}\sqrt{x^5} + c$

6 (a) $\frac{20}{3}$ or $6\frac{2}{3}$ (b) $\frac{33}{2}$ or $16\frac{1}{2}$ (c) $\frac{57}{8}$ or $7\frac{1}{8}$

(d) 6   (e) $\frac{10}{3}$ or $3\frac{1}{3}$ (f) 6

7 (a) $x^3 - \frac{3}{2}x^2 + c$

(b) $\displaystyle\int_1^2 3x(x-1)\,dx = \left[x^3 - \frac{3}{2}x^2\right]_1^2$

$= (8 - 6) - \left(1 - \frac{3}{2}\right)$

$= \frac{5}{2}$

8 (a) $\frac{5}{3}x^3 - 5x + c$ (b) $\frac{4}{3}x^3 + 6x^2 + 9x + c$

(c) $\frac{1}{3}x^3 + 2x - \dfrac{1}{x} + c$

9 7.527

10 $\frac{32}{3}$ or $10\frac{2}{3}$

11 (a) $\frac{1}{6}x^2 + \frac{4}{9}x^{\frac{3}{2}} + c$ or $\frac{1}{6}x^2 + \frac{4}{9}\sqrt{x^3} + c$

(b) $-\frac{1}{2}x^{-1} - \frac{1}{4}x^{-2} + c$ or $-\dfrac{1}{2x} - \dfrac{1}{4x^2} + c$

(c) $\frac{2}{3}x^{\frac{3}{2}} - 2x^{\frac{1}{2}} + c$ or $\frac{2}{3}\sqrt{x^3} - 2\sqrt{x} + c$

(d) $\frac{1}{5}x^{\frac{5}{2}} - 3x^{\frac{1}{2}} + c$ or $\frac{1}{5}\sqrt{x^5} - 3\sqrt{x} + c$

12 (a) $\sin\theta(1 - \cos^2\theta)$

$\equiv \sin\theta(\sin^2\theta)$

$\equiv \sin^3\theta$

(b) $\dfrac{5 - 5\sin^2 x}{\cos^2 x} \equiv \dfrac{5(1 - \sin^2 x)}{\cos^2 x}$

$\equiv \dfrac{5\cos^2 x}{\cos^2 x} \equiv 5$

(c) $\cos^2 t(\sec^2 t - 1)$

$\equiv \cos^2 t\left(\dfrac{1}{\cos^2 t} - 1\right)$

$\equiv 1 - \cos^2 t$

$\equiv \sin^2 t$

(d) $\sin 2\theta \tan\theta$

$\equiv 2\sin\theta\cos\theta\left(\dfrac{\sin\theta}{\cos\theta}\right)$

$\equiv 2\sin^2\theta$

(e) $\dfrac{\sin 2\theta}{\sin\theta} \equiv \dfrac{2\sin\theta\cos\theta}{\sin\theta} \equiv 2\cos\theta$

(f) $\dfrac{\cos 2t}{\cos^2 t} \equiv \dfrac{\cos^2 t - \sin^2 t}{\cos^2 t}$

$\equiv 1 - \dfrac{\sin^2 t}{\cos^2 t} \equiv 1 - \tan^2 t$

(g) $\dfrac{\cos 2\theta}{\sin^2\theta} \equiv \dfrac{1 - 2\sin^2\theta}{\sin^2\theta}$

$\equiv \dfrac{1}{\sin^2\theta} - 2$

$\equiv \operatorname{cosec}^2\theta - 2$

(h) $\dfrac{6\sin\theta\cos\theta}{\cos 2\theta} \equiv \dfrac{3\sin 2\theta}{\cos 2\theta}$

$\equiv 3\tan 2\theta$

13 (a) 0

(b) The area of region A is the same as the area of region B.

B Integrating e^x, $\sin x$, $\cos x$ and $\sec^2 x$ (p 58)

B1 (a) $-\sin x$ (b) $-\cos x + c$

B2 (a) $e^6 - e^2$ (b) $\left[\sin x\right]_0^{\frac{\pi}{2}} = 1$

B3 (a) $2\cos 2x$

     (b) (i) $\sin 2x + c$ (ii) $\frac{1}{2}\sin 2x + c$

B4 (a) $-5\sin 5x$

     (b) (i) $\cos 5x + c$ (ii) $-\frac{1}{5}\cos 5x + c$

B5 (a) $2e^{2x}$

     (b) (i) $e^{2x} + c$ (ii) $\frac{1}{2}e^{2x} + c$

B6 (a) $\frac{1}{4}e^{4x} + c$ (b) $-\frac{1}{6}\cos 6x + c$

     (c) $2\sin\frac{1}{2}x + c$

B7 (a) $4\sec^2 4x$ (b) $\frac{1}{4}\tan 4x + c$

B8 (a) $\frac{1}{5}(e^{0.5} - 1)$ (b) $\frac{1}{2}$

     (c) $\frac{1}{2}$ (d) 2

Exercise B (p 59)

1 (a) $\frac{1}{3}e^{3x} + c$

     (b) (i) $\frac{1}{3}(e^3 - 1)$ (ii) $\frac{1}{3}e^6 - \frac{7}{3}$

        (iii) $\frac{1}{3}(e^6 - e^3) - \frac{3}{2}$

2 (a) $\frac{1}{4}\sin 4x + c$

     (b) (i) $\frac{1}{4}$ (ii) $\frac{1}{4}\pi - \frac{1}{2}$

3 (a) $\frac{2}{3}\sin 3x - \frac{5}{2}e^{2x} + c$ (b) $\frac{4}{5}\tan 5x - \frac{5}{4}\cos 4x + c$

     (c) $-e^{-x} - \frac{1}{2}e^{-2x} + c$

4 (a) $\frac{1}{3}x^3 + \frac{1}{3}e^{3x} + c$ (b) $\frac{1}{2}x^2 + \frac{1}{4}\cos 4x + c$

     (c) $-e^{-x} - \frac{1}{x} + c$ (d) $-\frac{1}{5}\cos 5x - \frac{2}{3}x^{\frac{3}{2}} + c$

     (e) $\frac{1}{2}e^{2x} + \frac{1}{2}\tan 2x + c$ (f) $\frac{1}{3}\sin 3x + \dfrac{1}{2x^2} + c$

5 (a) $\left[\frac{3}{2}x^2 - \frac{1}{3}e^{3x}\right]_0^2 = \frac{19}{3} - \frac{1}{3}e^6$

     (b) $\left[\frac{1}{3}x^3 - 2\cos\frac{1}{2}x\right]_0^{\pi} = \frac{1}{3}\pi^3 + 2$

     (c) $\left[2x - 6\sin\frac{1}{2}x\right]_{-\pi}^{\pi} = 4\pi - 12$

     (d) $\left[-\frac{1}{2}e^{-2x} + x^2\right]_0^4 = \frac{33}{2} - \frac{1}{2}e^{-8}$

     (e) $\left[\frac{1}{4}x^4 + e^{-x}\right]_0^4 = 63 + e^{-4}$

     (f) $\left[x^2 + \frac{1}{2}\cos 2x\right]_0^{\frac{\pi}{2}} = \frac{1}{4}\pi^2 - 1$

6 (a) $\frac{1}{2}$ (b) 6

7 $\frac{26}{3} + \frac{1}{2}(e^{-6} - e^{-2})$

8 $\frac{14}{3} + 2e^2 - 2e^{\frac{1}{2}}$

9 $\int_0^k e^{-\frac{1}{4}x}\,dx = 1 \Rightarrow \left[-4e^{-\frac{1}{4}x}\right]_0^k = 1$

$\Rightarrow -4e^{-\frac{1}{4}k} + 4 = 1$

$\Rightarrow 4e^{-\frac{1}{4}k} = 3$

$\Rightarrow e^{-\frac{1}{4}k} = \frac{3}{4}$

$\Rightarrow \ln\left(e^{-\frac{1}{4}k}\right) = \ln\frac{3}{4}$

$\Rightarrow -\frac{1}{4}k = \ln\frac{3}{4}$

$\Rightarrow k = -4\ln\frac{3}{4}$

$\Rightarrow k = 4\ln\frac{4}{3}$

C Using trigonometric identities (p 60)

C1 (a) $\sin x\cos x = \frac{1}{2}\sin 2x$

     (b) $\int\frac{1}{2}\sin 2x\,dx = -\frac{1}{4}\cos 2x + c$

C2 (a) $\frac{1}{2}(\cos 2x + 1)$

     (b) $\int\frac{1}{2}(\cos 2x + 1)\,dx = \frac{1}{4}\sin 2x + \frac{1}{2}x + c$

C3 $\int\frac{1}{2}(1 - \cos 2x)\,dx = \frac{1}{2}x - \frac{1}{4}\sin 2x + c$

C4 (a) $\cos 2x = 1 - 2\sin^2 x$

Replace x by $\frac{1}{2}x$: $\cos x = 1 - 2\sin^2\frac{1}{2}x$

Rearrange: $2\sin^2\frac{1}{2}x = 1 - \cos x$

So $\sin^2\frac{1}{2}x = \frac{1}{2}(1 - \cos x)$

     (b) (i) $\int\frac{1}{2}(1 - \cos x)\,dx = \frac{1}{2}x - \frac{1}{2}\sin x + c$

        (ii) $\int_0^{\frac{\pi}{2}}\frac{1}{2}(1 - \cos x)\,dx = \frac{1}{2}\left[x - \sin x\right]_0^{\frac{\pi}{2}} = \frac{\pi}{4} - \frac{1}{2}$

Exercise C (p 60)

1 (a) $\frac{\pi}{2}$ (b) $\dfrac{\sqrt{3}}{8} + \dfrac{\pi}{12}$

     (c) $\dfrac{\sqrt{3}}{4} + \dfrac{\pi}{6}$

2 (a) $(1 + \cos x)^2 = 1 + 2\cos x + \cos^2 x$

$= 1 + 2\cos x + \frac{1}{2}(\cos 2x + 1)$

$= 1 + 2\cos x + \frac{1}{2}\cos 2x + \frac{1}{2}$

$= \frac{3}{2} + 2\cos x + \frac{1}{2}\cos 2x$

     (b) $\int(1 + \cos x)^2\,dx = \int\left(\frac{3}{2} + 2\cos x + \frac{1}{2}\cos 2x\right)dx$

$= \frac{3}{2}x + 2\sin x + \frac{1}{4}\sin 2x + c$

(c) $(1 + \sin x)^2 = 1 + 2\sin x + \sin^2 x$

$$= 1 + 2\sin x + \tfrac{1}{2}(1 - \cos 2x)$$
$$= \tfrac{3}{2} + 2\sin x - \tfrac{1}{2}\cos 2x$$
$$\int (1 + \sin x)^2 \, dx = \int \left(\tfrac{3}{2} + 2\sin x - \tfrac{1}{2}\cos 2x\right) dx$$
$$= \tfrac{3}{2}x - 2\cos x - \tfrac{1}{4}\sin 2x + c$$

3 (a) $(\cos\theta + \sin\theta)^2 = \cos^2\theta + 2\sin\theta\cos\theta + \sin^2\theta$

$$= (\cos^2 + \sin^2\theta) + 2\sin\theta\cos\theta$$
$$= 1 + \sin 2\theta$$
$$\int (\cos\theta + \sin\theta)^2 \, d\theta = \int (1 + \sin 2\theta)\, d\theta$$
$$= \theta - \tfrac{1}{2}\cos 2\theta + c$$

(b) $\int_0^{\frac{\pi}{2}} (\cos\theta - \sin\theta)^2 \, d\theta = \int_0^{\frac{\pi}{2}} (1 - \sin 2\theta)\, d\theta$

$$= \frac{\pi}{2} - 1$$

4 (a) $x - \tfrac{1}{4}\cos 4x + c$ **(b)** $\dfrac{\pi}{8} + \dfrac{1}{4}$

5 $\int_{\frac{\pi}{4}}^{\frac{\pi}{2}} \cos x (2\sin x - 1)\, dx$

$$= \int_{\frac{\pi}{4}}^{\frac{\pi}{2}} (2\sin x \cos x - \cos x)\, dx$$
$$= \int_{\frac{\pi}{4}}^{\frac{\pi}{2}} (\sin 2x - \cos x)\, dx$$
$$= \left[-\tfrac{1}{2}\cos 2x - \sin x\right]_{\frac{\pi}{4}}^{\frac{\pi}{2}}$$
$$= \left(-\tfrac{1}{2}\cos\pi - \sin\frac{\pi}{2}\right) - \left(-\tfrac{1}{2}\cos\frac{\pi}{2} - \sin\frac{\pi}{4}\right)$$
$$= \left(\tfrac{1}{2} - 1\right) - \left(0 - \frac{1}{\sqrt{2}}\right)$$
$$= \frac{1}{\sqrt{2}} - \frac{1}{2}$$
$$= \frac{\sqrt{2}}{2} - \frac{1}{2} = \tfrac{1}{2}(\sqrt{2} - 1) \quad \text{as required}$$

6 (a) $\int \tan^2 x \, dx = \int (\sec^2 x - 1)\, dx$
$$= \tan x - x + c$$

(b) $3\sqrt{3} - \pi$

7 (a) $2\tan\tfrac{1}{2}x - x + c$

(b) $\int_0^{\frac{\pi}{2}} \tan^2 \tfrac{1}{2}x \, dx$

$$= \left[2\tan\tfrac{1}{2}x - x\right]_0^{\frac{\pi}{2}} \quad \text{from part (a)}$$
$$= \left(2\tan\frac{\pi}{4} - \frac{\pi}{2}\right) - (2\tan 0 - 0)$$
$$= 2 - \frac{\pi}{2} \quad \text{as required}$$

8 $\dfrac{1}{2} + \dfrac{\pi}{4}$

9 (a) $\tfrac{1}{12}\sin 6x + \tfrac{1}{2}x + c$ **(b)** $\dfrac{\pi}{6}$

10 $\dfrac{\pi}{2} - \sqrt{2}$

D Integrating $\dfrac{1}{x}$ (p 61)

D1 (a) $\ln 5$ **(b)** $\ln\tfrac{1}{2}$ **(c)** $9 + \ln\tfrac{5}{2}$

D2 $\tfrac{1}{3}\ln|x| + c \;\; (x \neq 0)$

D3 (a) $\dfrac{2}{2x+1}$

(b) $\int \dfrac{1}{2x+1}\, dx = \tfrac{1}{2}\int \dfrac{2}{2x+1}\, dx$
$$= \tfrac{1}{2}\ln|2x+1| + c \;\; \left(x \neq -\tfrac{1}{2}\right)$$

D4 (a) $\dfrac{5}{5x-2}$ **(b)** $\tfrac{1}{5}\ln|5x-2| + c \;\; \left(x \neq \tfrac{2}{5}\right)$

Exercise D (p 63)

1 (a) $x + \tfrac{1}{2}x^2 - \ln|x| + c$

(b) $\tfrac{1}{2}e^{2x} + 2\ln|x| + c$

(c) $-\tfrac{1}{6}\cos 6x - \tfrac{1}{2}\ln|x| + c$

(d) $\ln|x| + \tfrac{2}{3}x^{\frac{3}{2}} + c$

(e) $\ln|x| + 2\sqrt{x} + c$

(f) $\tfrac{1}{4}\sin 4x + \tfrac{1}{4}\ln|x| + c$

(g) $5\ln|x| - \tfrac{1}{3}e^{-3x} + c$

(h) $x + \ln|x| - \dfrac{1}{x} - \dfrac{1}{2x^2} + c$

(i) $\tfrac{1}{4}\ln|4x - 1| + c$

2 (a) $4\ln 3$

(b) $\int_2^{2\sqrt{3}} \dfrac{4}{x}\, dx = 4\ln\sqrt{3} = 2\ln 3$
$$= \text{half the area in part (a) as required}$$

3 (a) $\frac{1}{2}\ln 3$

(b) $\frac{14}{3} - \ln 4$

(c) $\frac{1}{3}(e^{-3} - e^{-6} - \ln 2)$

(d) $168 + 4\ln 4$

(e) $2 + 3\ln\frac{2}{3}$ or $2 - 3\ln\frac{3}{2}$

(f) $\frac{1}{3}\ln\frac{5}{2}$

4 (a) $\dfrac{1}{x^2} + \dfrac{1}{x}$ **(b)** $\frac{2}{3} + \ln 3$

5 $\frac{4}{9} + \ln 3$

6 (a) $\frac{1}{3}$ **(b)** $4 - 2\ln 3$

E Integration by substitution (p 64)

E1 $\frac{1}{2}(x^3 - 1)^2 + c$

E2 $-\frac{1}{4}(5 - x^2)^4 + c$

E3 $-\dfrac{1}{2(5x+1)^2} + c$ or $-\frac{1}{2}(5x+1)^{-2} + c$

E4 $\frac{1}{10}(x^2 - 3)^5 + c$

E5 $\frac{1}{16}(x^4 - 1)^4 + c$

E6 $\frac{1}{14}(2x+1)^7 - \frac{1}{12}(2x+1)^6 + c$

E7 $\frac{2}{15}(3x - 2)^5 + \frac{1}{3}(3x - 2)^4 + c$

E8 $\frac{1}{24}(2x+3)^6 - \frac{3}{20}(2x+3)^5 + c$

Exercise E (p 66)

1 $-\dfrac{1}{3(2x-7)^3} + c$ or $-\frac{1}{3}(2x-7)^{-3} + c$

2 (a) $\frac{1}{5}(x^3 - 1)^5 + c$ **(b)** $-\dfrac{1}{3(x^3-1)} + c$

(c) $\frac{1}{3}(x^2 - 4)^{\frac{3}{2}} + c$ **(d)** $\frac{1}{24}(2x^2 + 1)^6 + c$

3 (a) $\frac{1}{12}(2x - 3)^6 + \frac{3}{10}(2x - 3)^5 + c$

(b) $\frac{1}{24}(2x+1)^6 - \frac{1}{20}(2x+1)^5 + c$

(c) $\frac{2}{45}(3x - 2)^{\frac{5}{2}} + \frac{4}{27}(3x - 2)^{\frac{3}{2}} + c$

(d) $\frac{1}{5}(5x - 2) + \frac{2}{5}\ln|5x - 2| + c$

(e) $\frac{1}{16}(4x + 1) - \frac{1}{16}\ln|4x + 1| + c$

(f) $-\frac{1}{4}(2x - 1)^{-1} - \frac{1}{8}(2x - 1)^{-2} + c$

4 (a) $u = x^2 + 3$, $\frac{1}{4}(x^2 + 3)^4 + c$

(b) $u = x^2 + 5$, $\frac{1}{3}(x^2 + 5)^{\frac{3}{2}} + c$

(c) $u = 6x + 5$, $\frac{1}{252}(6x + 5)^7 - \frac{5}{216}(6x + 5)^6 + c$

(d) $u = 1 - 2x$, $\frac{1}{24}(1 - 2x)^6 - \frac{1}{20}(1 - 2x)^5 + c$

(e) $u = x^2 - 1$, $\sqrt{x^2 - 1} + c$

(f) $u = x - 1$, $\frac{2}{3}(x - 1)^{\frac{3}{2}} + 2(x - 1)^{\frac{1}{2}} + c$

F Further integration by substitution (p 66)

F1 $\left(\dfrac{u-1}{2}\right) \times \sqrt{u} \times \frac{1}{2}$

$= \frac{1}{4}u^{\frac{1}{2}}(u - 1) = \frac{1}{4}\left(u^{\frac{3}{2}} - u^{\frac{1}{2}}\right)$

So the integral is $\frac{1}{4}\int_1^9 \left(u^{\frac{3}{2}} - u^{\frac{1}{2}}\right)du$

and its value is $\frac{298}{15}$ or $19\frac{13}{15}$.

F2 $\frac{8}{3}$ or $2\frac{2}{3}$

F3 $\frac{98}{3}$ or $32\frac{2}{3}$

F4 $\dfrac{\pi}{2}$

F5 For $x = \tan\theta$ we get $\dfrac{dx}{d\theta} = \sec^2\theta$.

So $dx = \sec^2\theta\, d\theta$

When $x = 0$, $\theta = 0$ and when $x = 1$, $\theta = \dfrac{\pi}{4}$

So $\int_0^1 \dfrac{1}{1+x^2}\, dx = \int_0^{\frac{\pi}{4}} \dfrac{1}{1 + \tan^2\theta}\sec^2\theta\, d\theta$

$= \int_0^{\frac{\pi}{4}} \dfrac{\sec^2\theta}{\sec^2\theta}\, d\theta$

$= \int_0^{\frac{\pi}{4}} 1\, d\theta$ as required

Its value is $\dfrac{\pi}{4}$.

F6 For $x = \tan\theta$ we get $\dfrac{dx}{d\theta} = \sec^2\theta$.

So $dx = \sec^2\theta\, d\theta$

When $x = 0$, $\theta = 0$ and when $x = 1$, $\theta = \dfrac{\pi}{4}$

So $\int_0^1 \dfrac{1}{\left(1+x^2\right)^2}\, dx = \int_0^{\frac{\pi}{4}} \dfrac{1}{\left(1 + \tan^2\theta\right)^2}\sec^2\theta\, d\theta$

$= \int_0^{\frac{\pi}{4}} \dfrac{\sec^2\theta}{\left(\sec^2\theta\right)^2}\, d\theta$

$= \int_0^{\frac{\pi}{4}} \dfrac{1}{\sec^2\theta}\, d\theta$

$= \int_0^{\frac{\pi}{4}} \cos^2\theta\, d\theta$ as required

Its value is $\dfrac{\pi}{8} + \dfrac{1}{4}$.

F7 (a) From $x = \sin\theta$ we get $\dfrac{dx}{d\theta} = \cos\theta$.

So $dx = \cos\theta\, d\theta$

When $x = 0$, $\theta = 0$ and when $x = 1$, $\theta = \tfrac{1}{2}\pi$

So $\displaystyle\int_0^1 x^2\sqrt{1-x^2}\, dx$

$\displaystyle = \int_0^{\frac{1}{2}\pi} \sin^2\theta\sqrt{1-\sin^2\theta}\,\cos\theta\, d\theta$

$\displaystyle = \int_0^{\frac{1}{2}\pi} \sin^2\theta\cos^2\theta\, d\theta$ as required

(b) $\sin^2\theta\cos^2\theta = (\sin\theta\cos\theta)^2$

$\qquad = \left(\tfrac{1}{2}\times 2\sin\theta\cos\theta\right)^2$

$\qquad = \left(\tfrac{1}{2}\sin 2\theta\right)^2$

$\qquad = \tfrac{1}{4}\sin^2 2\theta$ as required

(c) $\dfrac{\pi}{16}$

Exercise F (p 68)

1 (a) $\dfrac{5823}{10}$ or $582\tfrac{3}{10}$

(b) $\dfrac{1}{3}$

(c) $\dfrac{144}{5}$ or $28\tfrac{4}{5}$

(d) $\tfrac{1}{2}(e-1)$

(e) $\dfrac{\pi}{2}$

(f) $\dfrac{2}{3}$

(g) $\dfrac{2}{3}\sqrt{3} - \dfrac{\pi}{6}$ or $\dfrac{2}{\sqrt{3}} - \dfrac{\pi}{6}$

(h) $\dfrac{8}{3}$ or $2\tfrac{2}{3}$

2 $u = 1 + \sin\theta$ gives $\dfrac{du}{d\theta} = \cos\theta$.

So $du = \cos\theta\, d\theta$

When $\theta = 0$, $u = 1 + \sin 0 = 1$ and

when $\theta = \dfrac{\pi}{2}$, $u = 1 + \sin\dfrac{\pi}{2} = 2$

So $\displaystyle\int_0^{\frac{\pi}{2}} \dfrac{\sin 2\theta}{1+\sin\theta}\, d\theta = \int_0^{\frac{\pi}{2}} \dfrac{2\sin\theta\cos\theta}{1+\sin\theta}\, d\theta$

$\displaystyle = \int_1^2 \dfrac{2(u-1)}{u}\, du$

$\displaystyle = 2\int_1^2\left(1 - \dfrac{1}{u}\right)du$

$= 2\big[u - \ln u\big]_1^2$

$= 2\big((2 - \ln 2) - (1 - \ln 1)\big)$

$= 2(1 - \ln 2)$ as required

G Integrating $\dfrac{f'(x)}{f(x)}$ (p 68)

G1 $\ln|x^2 + 1| + c$

G2 $\ln|x^3 - 1| + c$

G3 $\ln|x^4 + 3x^2 - 5| + c$

Exercise G (p 69)

1 (a) $\ln|x^3 - 1| + c$

(b) $\ln|x^2 + x| + c$

(c) $\tfrac{1}{5}\ln|x^5 + 1| + c$

(d) $\tfrac{1}{2}\ln|x^2 - 2x| + c$

(e) $\tfrac{1}{4}\ln|2x^2 + 4x + 1| + c$

(f) $\tfrac{1}{3}\ln|3e^x - 1| + c$

2 (a) $-\tfrac{1}{2}\ln|4 - x^2| + c$

(b) $\left[-\tfrac{1}{2}\ln|4 - x^2|\right]_0^1 = -\tfrac{1}{2}\ln 3 + \tfrac{1}{2}\ln 4 = \tfrac{1}{2}\ln\left(\tfrac{4}{3}\right)$

3 (a) Area required

$\displaystyle = \int_0^2 \dfrac{x}{x^2+1}\, dx = \tfrac{1}{2}\int_0^2 \dfrac{2x}{x^2+1}\, dx = \left[\tfrac{1}{2}\ln|x^2+1|\right]_0^2$

$= \tfrac{1}{2}\ln 5$

(b) $\tfrac{1}{3}\ln 9$

4 (a) $\dfrac{\sin x}{\cos x}$

(b) The derivative of $\cos x$ is $-\sin x$, so we have

$-\dfrac{f'(x)}{f(x)}$ where $f(x) = \cos x$.

(c) $\ln|\sin x| + c$

5 (a) $\tfrac{1}{3}\ln|e^{3x} + 1| + c$

(b) $\displaystyle\int_0^1 \dfrac{e^{3x}}{e^{3x}+1}\, dx = \left[\tfrac{1}{3}\ln|e^{3x}+1|\right]_0^1$

$= \tfrac{1}{3}\ln(e^3 + 1) - \tfrac{1}{3}\ln(e^0 + 1)$

$= \tfrac{1}{3}\big(\ln(e^3 + 1) - \ln 2\big)$

$= \tfrac{1}{3}\ln\left(\dfrac{e^3 + 1}{2}\right)$

6 (a) $\dfrac{1}{2\sqrt{x}}$　　　　**(b)** $2\ln|1 + \sqrt{x}| + c$

7 $\int_0^{\frac{\pi}{2}} \dfrac{\cos\theta}{3+\sin\theta}\,d\theta$

$= [\ln(3+\sin\theta)]_0^{\frac{\pi}{2}}$

$= \ln\left(3+\sin\dfrac{\pi}{2}\right) - \ln(3+\sin 0)$

$= \ln 4 - \ln 3$

$= \ln\frac{4}{3}$ as required

H Integration by parts (p 70)

H1 (a) The derivative of $x\cos x$ is $-x\sin x + \cos x$
so $\int(-x\sin x + \cos x)\,dx = x\cos x + c$.

(b) $\int x\sin x\,dx = \int\cos x\,dx - x\cos x - c$
$= \sin x - x\cos x + c$
Note that as c is an arbitrary constant, we can choose that it be added rather than subtracted without altering the effect.

H2 (a) The derivative of xe^x is $xe^x + e^x$
so $\int(xe^x + e^x)\,dx = xe^x + c$.

(b) $\int xe^x\,dx = xe^x - \int e^x\,dx + c$
$= xe^x - e^x + c$

H3 (a) The derivative of $x\sin 2x$ is $2x\cos 2x + \sin 2x$
so $\int(2x\cos 2x + \sin 2x)\,dx = x\sin 2x + c$.

(b) $\int 2x\cos 2x\,dx = x\sin 2x - \int\sin 2x\,dx + c$
$= x\sin 2x + \frac{1}{2}\cos 2x + c$

H4 $-\frac{1}{4}x\cos 4x + \frac{1}{16}\sin 4x + c$

H5 $\frac{1}{4}xe^{4x} - \frac{1}{16}e^{4x} + c$

H6 (a) $\frac{1}{4}x\sin 4x + \frac{1}{16}\cos 4x + c$

(b) $\frac{1}{5}xe^{5x} - \frac{1}{25}e^{5x} + c$

(c) $-\frac{1}{4}xe^{-4x} - \frac{1}{16}e^{-4x} + c$

H7 The integral to be found is harder than the original, not easier.

H8 (a) We do not yet know how to integrate $\ln x$.

(b) $\frac{1}{2}x^2\ln x - \frac{1}{4}x^2 + c$

Exercise H (p 72)

1 (a) $\frac{2}{7}x\sin 7x + \frac{2}{49}\cos 7x + c$

(b) $\frac{1}{2}xe^{6x} - \frac{1}{12}e^{6x} + c$

(c) $-2x\cos\frac{1}{2}x + 4\sin\frac{1}{2}x + c$

(d) $-xe^{-x} - e^{-x} + c$

(e) $-\frac{1}{3}xe^{-3x} - \frac{1}{9}e^{-3x} + c$

(f) $-2xe^{-\frac{1}{2}x} - 4e^{-\frac{1}{2}x} + c$

2 (a) $\dfrac{\pi}{2} - 1$ **(b)** π **(c)** $3e^4 + 1$

(d) $\frac{1}{9}$ **(e)** $2 - \dfrac{4}{e}$ **(f)** $8e^3 + 4$

3 (a) $0, \dfrac{\pi}{2}, \pi$ **(b)** $A = \dfrac{\pi}{4}, B = -\frac{3}{4}\pi$

4 $\frac{81}{4}\ln 3 - 5$

5 (a) $x\ln x - x + c$ **(b)** $5\ln 5 - 4$

6 2.68

I Further integration by parts (p 73)

I1 (a) $xe^x - e^x + k$ where k is a constant

(b) Let $u = x^2$ and $\dfrac{dv}{dx} = e^x$

Then $v = \int e^x\,dx = e^x$ and $\dfrac{du}{dx} = 2x$

So $\int x^2 e^x\,dx = x^2 e^x - \int 2xe^x\,dx$
$= x^2 e^x - 2\int xe^x\,dx$
$= x^2 e^x - 2(xe^x - e^x + k)$
$= x^2 e^x - 2xe^x + 2e^x - 2k$
$= e^x(x^2 - 2x + 2) + c$ where c is a constant

I2 (a) Let $u = x^2$ and $\dfrac{dv}{dx} = \cos x$

So $v = \int\cos x\,dx = \sin x$ and $\dfrac{du}{dx} = 2x$

Hence $\int x^2\cos x\,dx = x^2\sin x - \int 2x\sin x\,dx$
$= x^2\sin x - 2\int x\sin x\,dx$

(b) $-x\cos x + \sin x + k$ where k is a constant

(c) $x^2\sin x + 2x\cos x - 2\sin x + c$

I3 $\frac{1}{4}x^2 e^{4x} - \frac{1}{8}xe^{4x} + \frac{1}{32}e^{4x} + c$ or $\frac{1}{32}e^{4x}(8x^2 - 4x + 1) +$

I4 (a) Let $u = x^2$ and $\dfrac{dv}{dx} = \sin 2x$

So $v = \int\sin 2x\,dx = -\frac{1}{2}\cos 2x$ and $\dfrac{du}{dx} = 2x$

Hence $\int_0^{\frac{\pi}{4}} x^2\sin 2x\,dx$

$= \left[-\frac{1}{2}x^2\cos 2x\right]_0^{\frac{\pi}{4}} - \int_0^{\frac{\pi}{4}}\left(-\frac{1}{2}\cos 2x\right)2x\,dx$

$= \left[-\frac{1}{2}x^2\cos 2x\right]_0^{\frac{\pi}{4}} + \int_0^{\frac{\pi}{4}} x\cos 2x\,dx$

(b) Let $u = x$ and $\dfrac{dv}{dx} = \cos 2x$

So $v = \int \cos 2x \, dx = \tfrac{1}{2} \sin 2x$ and $\dfrac{du}{dx} = 1$

Hence $\int_0^{\frac{\pi}{4}} x \cos 2x \, dx$

$= \left[\tfrac{1}{2} x \sin 2x \right]_0^{\frac{\pi}{4}} - \int_0^{\frac{\pi}{4}} \tfrac{1}{2} \sin 2x \, dx$

$= \left(\dfrac{\pi}{8} \sin \dfrac{\pi}{2} - 0 \right) + \left[\tfrac{1}{4} \cos 2x \right]_0^{\frac{\pi}{4}}$

$= \dfrac{\pi}{8} + \left(\tfrac{1}{4} \cos \dfrac{\pi}{2} - \tfrac{1}{4} \cos 0 \right)$

$= \dfrac{\pi}{8} - \dfrac{1}{4}$ as required

(c) $\dfrac{\pi}{8} - \dfrac{1}{4}$

Exercise I (p 74)

1 (a) $-2x^2 \cos x + 4x \sin x + 4 \cos x + c$

(b) $\tfrac{1}{4} x^2 \sin 2x + \tfrac{1}{4} x \cos 2x - \tfrac{1}{8} \sin 2x + c$

(c) $-x^2 e^{-x} - 2x e^{-x} - 2 e^{-x} + c$ or
$-e^{-x}(x^2 + 2x + 2) + c$

2 (a) $\tfrac{1}{2} x^2 e^{2x} - \tfrac{1}{2} x e^{2x} + \tfrac{1}{4} e^{2x} + c$ or
$\tfrac{1}{4} e^{2x}(2x^2 - 2x + 1) + c$

(b) $\int_0^2 x^2 e^{2x} \, dx = \left[\tfrac{1}{4} e^{2x}(2x^2 - 2x + 1) \right]_0^2$

$= \left(\tfrac{1}{4} e^4 \times 5 \right) - \left(\tfrac{1}{4} e^0 \times 1 \right)$

$= \tfrac{5}{4} e^4 - \tfrac{1}{4}$

$= \tfrac{1}{4} \left(5 e^4 - 1 \right)$ as required

3 $\pi^2 - 8$

4 $\tfrac{1}{9} (\pi^2 - 4)$

Mixed questions (p 75)

1 (a) Area $= \tfrac{1}{2}$ base $\times$ height $= \tfrac{1}{2} \times \tfrac{5}{12} \pi \times \sin \tfrac{5}{6} \pi$

$= \dfrac{5}{48} \pi$

(b) $\dfrac{\sqrt{3}}{4} + \tfrac{1}{2} - \tfrac{5}{48} \pi$

2 (a)

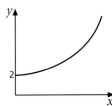

(b) At A, $y = 2 e^{\frac{1}{2}x} = 6 \Rightarrow e^{\frac{1}{2}x} = 3 \Rightarrow \tfrac{1}{2} x = \ln 3$
$\Rightarrow x = 2 \ln 3$

(c) $12 \ln 3 - 8$

3 (a) $-\tfrac{1}{3} \cos (3x - 1) + c$

(b) $\tfrac{1}{6} \ln |6x - 1| + c$

(c) $2 \tan \tfrac{1}{2} x + c$

4 (a) $-\dfrac{1}{4(4 + 3x)^4} + c$ or $-\tfrac{1}{4}(4 + 3x)^{-4} + c$

(b) $\tfrac{1}{2} \ln |x^2 - 5| + c$

(c) $-\dfrac{2}{x - 5} + c$ or $\dfrac{2}{5 - x} + c$

5 (a) $\tfrac{3}{2}$ **(b)** $\dfrac{\pi^2}{2} + 2$ **(c)** $\tfrac{2}{3} \ln \tfrac{26}{7}$

6 (a) $e^a = 2a + 1$

(b) Area under $y = 2x + 1$ is $a^2 + a$
(using trapezium formula)
Area under $y = e^x$ is $e^a - 1 = 2a$
(using integration)
Area required $= a^2 - a$

7 (a) $\cos x = \cos^2 \tfrac{1}{2} x - \sin^2 \tfrac{1}{2} x$
$= \cos^2 \tfrac{1}{2} x - \left(1 - \cos^2 \tfrac{1}{2} x \right)$
$= 2 \cos^2 \tfrac{1}{2} x - 1$
$\Rightarrow \cos x + 1 = 2 \cos^2 \tfrac{1}{2} x$
$\Rightarrow \cos^2 \tfrac{1}{2} x = \tfrac{1}{2}(\cos x + 1)$ as required

(b) $\tfrac{1}{2} \pi$

8 $\int_0^{\frac{\pi}{3}} \tan^2 x \, dx = \int_0^{\frac{\pi}{3}} \left(\sec^2 x - 1 \right) dx$

$= \left[\tan x - x \right]_0^{\frac{\pi}{3}}$

$= \left(\tan \dfrac{\pi}{3} - \dfrac{\pi}{3} \right) - (\tan 0 - 0)$

$= \sqrt{3} - \dfrac{\pi}{3}$ as required

9 (a) $\tfrac{1}{16} \sin 4x - \tfrac{1}{4} x \cos 4x + c$

(b) $\tfrac{1}{5} x^5 \ln x - \tfrac{1}{25} x^5 + c$ or $\tfrac{1}{25} x^5 (5 \ln x - 1) + c$

(c) $\tfrac{1}{3} x^2 e^{3x} - \tfrac{2}{9} x e^{3x} + \tfrac{2}{27} e^{3x} + c$ or
$\tfrac{1}{27} e^{3x}(9x^2 - 6x + 2) + c$

10 $\int_1^2 \frac{x+1}{x^2+2x+1}\,dx = \frac{1}{2}\int_1^2 \frac{2x+2}{x^2+2x+1}\,dx$

$= \frac{1}{2}\big[\ln(x^2+2x+1)\big]_1^2$

$= \frac{1}{2}(\ln 9 - \ln 4)$

$= \frac{1}{2}\ln\frac{9}{4}$

$= \ln\left(\frac{9}{4}\right)^{\frac{1}{2}} = \ln\frac{3}{2}$ as required

11 (a) $\frac{3}{2}$ **(b)** $3 - 2\ln 2$

12 $\frac{1}{2}\ln 2$ or $\ln\sqrt{2}$

13 $\frac{1}{3}e^{x^3} + c$

14 $\frac{64}{3}\ln 2 - \frac{7}{2}$

15 $\frac{1}{6}\arctan\left(\frac{3x}{2}\right) + c$

16 $\frac{\pi}{6} - \frac{\sqrt{3}}{8}$

17 (a) $-\frac{1}{2}xe^{-2x} - \frac{1}{4}e^{-2x} + c$ or $-\frac{1}{4}e^{-2x}(2x+1) + c$

(b) $\frac{1}{6}(x-2)^6 + \frac{2}{5}(x-2)^5 + c$

(c) $3\ln|1 - \cos\theta| + c$

(d) $-\frac{1}{x}\ln|x| - \frac{1}{x} + c$ or $-\frac{1}{x}(\ln|x| + 1) + c$

(e) $2e^{(x^2+5)} + c$

(f) $5x^2e^{\frac{1}{5}x} - 50xe^{\frac{1}{5}x} + 250e^{\frac{1}{5}x} + c$ or

$5e^{\frac{1}{5}x}(x^2 - 10x + 50) + c$

Test yourself (p 77)

1 (a) $\frac{1}{6}e^{6x} + c$ **(b)** $4\sin\frac{1}{4}x + c$

(c) $-\frac{1}{2}\cos(2x+1) + c$

2 (a) $6\ln 3\frac{5}{2}$ **(b)** $\frac{1}{2}\left(1 - \frac{1}{e}\right)$ **(c)** $\frac{1}{3}\ln 7$

(d) $\frac{1}{2}e^2 - \frac{3}{2}$ **(e)** 1

3 $\ln\frac{7}{3}$

4 (a) $\tan\theta + 4\theta + c$ **(b)** $-\frac{1}{4}\cos 2x + c$

(c) $3\tan t + c$

5 $-\frac{1}{4}(2x+1)\cos 4x + \frac{1}{8}\sin 4x + c$

6 $\frac{2}{5}(x+1) - \frac{4}{3}(x+1)^{\frac{3}{2}} + c$

7 4

8 $u = 2 - \sin x$ gives $du = -\cos x\,dx$

$\int \frac{\sin 2x}{(2-\sin x)^3}\,dx = \int \frac{2\sin x\cos x}{(2-\sin x)^3}\,dx$

$= \int \frac{-2\sin x}{(2-\sin x)^3}(-\cos x\,dx)$

$\int \frac{-2(2-u)}{u^3}\,du = \int \frac{2u-4}{u^3}\,du$

$= \int (2u^{-2} - 4u^{-3})\,du$

$= \frac{2u^{-1}}{-1} - \frac{4u^{-2}}{-2} + c$

$= -2u^{-1} + 2u^{-2} + c$

$= 2\left(\frac{1}{u^2} - \frac{1}{u}\right) + c = 2\left(\frac{1-u}{u^2}\right) + c$

$= 2\left(\frac{1-(2-\sin x)}{(2-\sin x)^2}\right) + c$

$= \frac{2(\sin x - 1)}{(2-\sin x)^2} + c$ as required

9 (a) $9\ln 3 - \frac{26}{9}$

(b) $x = \sin\theta$ gives $dx = \cos\theta\,d\theta$

So $\int \frac{1}{(1-x^2)^{\frac{3}{2}}}\,dx = \int \frac{1}{(1-\sin^2\theta)^{\frac{3}{2}}}\cos\theta\,d\theta$

$= \int \frac{\cos\theta}{\left(\sqrt{1-\sin^2\theta}\right)^3}\,d\theta$

$= \int \frac{\cos\theta}{\cos^3\theta}\,d\theta$

$= \int \frac{1}{\cos^2\theta}\,d\theta$

$= \int \sec^2\theta\,d\theta$

$= \tan\theta + c$

$= \frac{\sin\theta}{\cos\theta} + c$

$= \frac{\sin\theta}{\sqrt{1-\sin^2\theta}} + c$

$= \frac{x}{\sqrt{1-x^2}} + c$

$= \frac{x}{(1-x^2)^{\frac{1}{2}}} + c$ as required

10 $-8x^2\cos\frac{1}{4}x + 64x\sin\frac{1}{4}x + 256\cos\frac{1}{4}x + c$

11 $x = 2\sqrt{3}\sin\theta$ gives $x^2 = 12\sin^2\theta$

and $dx = 2\sqrt{3}\cos\theta\,d\theta$

When $x = \sqrt{6}, \sin\theta = \dfrac{\sqrt{6}}{2\sqrt{3}} = \dfrac{1}{\sqrt{2}}$ so $\theta = \dfrac{\pi}{4}$

and when $x = 2\sqrt{3}, \sin\theta = 1$ so $\theta = \dfrac{\pi}{2}$.

Hence $\displaystyle\int_{\sqrt{6}}^{2\sqrt{3}} \sqrt{12 - x^2}\,dx$

$= \displaystyle\int_{\frac{\pi}{4}}^{\frac{\pi}{2}} \sqrt{12 - 12\sin^2\theta}\,(2\sqrt{3}\cos\theta\,d\theta)$

$= \displaystyle\int_{\frac{\pi}{4}}^{\frac{\pi}{2}} \sqrt{12}\sqrt{1 - \sin^2\theta}\,(2\sqrt{3}\cos\theta\,d\theta)$

$= \displaystyle\int_{\frac{\pi}{4}}^{\frac{\pi}{2}} 12\cos^2\theta\,d\theta$

As $\cos^2\theta = \frac{1}{2}(\cos 2\theta + 1)$,

$\displaystyle\int_{\frac{\pi}{4}}^{\frac{\pi}{2}} 12\cos^2\theta\,d\theta = 6\int_{\frac{\pi}{4}}^{\frac{\pi}{2}} (\cos 2\theta + 1)\,d\theta$

$= 6\left[\frac{1}{2}\sin 2\theta + \theta\right]_{\frac{\pi}{4}}^{\frac{\pi}{2}}$

$= 6\left(\left(\frac{1}{2}\sin\pi + \frac{\pi}{2}\right) - \left(\frac{1}{2}\sin\frac{\pi}{2} + \frac{\pi}{4}\right)\right)$

$= 6\left(\frac{\pi}{2} - \frac{1}{2} - \frac{\pi}{4}\right)$

$= 6\left(\frac{\pi}{4} - \frac{1}{2}\right)$

$= 3\left(\frac{\pi}{2} - 1\right)$ as required

12 $\dfrac{\pi}{6} - \dfrac{1}{3}$

6 Integration 2

A Area under a curve defined parametrically
(p 78)

A1 1.75

A2 $10\frac{2}{3}$

A3 $\frac{15}{8} - 2\ln 2$

Exercise A (p 79)

1 8.25

2 (a) 152   **(b)** $\frac{8}{15}$ **(c)** $2 - \ln 2$

   (d) $\frac{1}{2} - \frac{\pi}{6}$ **(e)** $2\sqrt{2} - 4$ **(f)** $2(e^2 + 1)$

3 (a) (i) 0   **(ii)** π

   (b) 3π

B Volume of a solid of revolution about the x-axis (p 80)

B1 (a) Cone   **(b)** $\frac{1}{4}x^2$ **(c)** $\dfrac{9\pi}{4}$

B2 (a) The gradient m is $\dfrac{r}{h}$.

The line passes through the origin so the equation is $y = \dfrac{r}{h}x$.

(b) $y^2 = \dfrac{r^2}{h^2}x^2$

$V = \displaystyle\int_0^h \pi\frac{r^2}{h^2}x^2\,dx = \pi\frac{r^2}{h^2}\left[\frac{1}{3}x^3\right]_0^h$

$= \pi\dfrac{r^2}{h^2}\left(\frac{1}{3}h^3 - 0\right) = \frac{1}{3}\pi r^2 h$

B3 (a) x^4 **(b)** $\frac{31}{5}\pi$

B4 (a) Sphere

(b) $\displaystyle\int_{-r}^{r} \pi y^2\,dx = \pi\int_{-r}^{r}(r^2 - x^2)\,dx = \pi\left[r^2 x - \frac{1}{3}x^3\right]_{-r}^{r}$

$= \pi\left[\left(r^3 - \frac{1}{3}r^3\right) - \left(r^2(-r) - \frac{1}{3}(-r)^3\right)\right]$

$= \pi\left(r^3 - \frac{1}{3}r^3 + r^3 - \frac{1}{3}r^3\right)$

$= \frac{4}{3}\pi r^3$

Exercise B (p 82)

1 $\dfrac{32\pi}{3}$

2 $\dfrac{8\pi}{15}$

3 (a) $\frac{1}{2}\pi(e^6-1)$ **(b)** $\frac{21\pi}{64}$ **(c)** $\frac{40\pi}{3}$

(d) $\frac{2\pi}{3}$ **(e)** $\frac{1}{4}\pi^2$ **(f)** $\frac{\pi(e^2-3)}{e^2}$

(g) $\frac{\pi}{2}\ln 3$ **(h)** $\frac{1}{2}\pi\ln 10$

C Solid of revolution defined parametrically

(p 83)

C1 $\pi\int_0^1 (1+2t)^2 2t\,dt = \pi\int_0^1 (2t+8t^2+8t^3)\,dt$

$= \pi\left[t^2 + \frac{8}{3}t^3 + 2t^4\right]_0^1 = \pi(1+\frac{8}{3}+2) = \frac{17\pi}{3}$

Exercise C (p 83)

1 24π

2 (a) $\frac{9\pi}{20}$ **(b)** $\frac{\pi}{20}(4e^5+5e^4-9)$

(c) $\frac{\pi}{15}$ **(d)** $\frac{1}{2}\pi(1+\frac{1}{2}\pi)$

3 $\frac{1}{12}\pi(44-39\sqrt{2}+24\ln 2)$

4 $\frac{1}{2}\pi(3e^4+1)$

5 $\pi\ln\sqrt{2} = \frac{1}{2}\pi\ln 2$

6 (a) $\frac{d}{dx}(\ln(\sec x+\tan x)) = \dfrac{\frac{d}{dx}(\sec x+\tan x)}{\sec x+\tan x}$

$= \dfrac{\sec x\tan x+\sec^2 x}{\sec x+\tan x}$

$= \dfrac{\sec x(\tan x+\sec x)}{\sec x+\tan x}$

$= \sec x$

(b) $\frac{dx}{d\theta} = \cos\theta$ so $dx = \cos\theta\,d\theta$

$y^2 = \tan^2\theta$

$V = \pi\int y^2\,dx = \pi\int_0^{\frac{1}{4}\pi}\tan^2\theta\cos\theta\,d\theta$

$= \pi\int_0^{\frac{1}{4}\pi}\frac{\sin^2\theta}{\cos\theta}\cos\theta\,d\theta = \pi\int_0^{\frac{1}{4}\pi}\frac{\sin^2\theta}{\cos\theta}\,d\theta$

$= \pi\int_0^{\frac{1}{4}\pi}\frac{(1-\cos^2\theta)}{\cos\theta}\,d\theta = \pi\int_0^{\frac{1}{4}\pi}\left(\frac{1}{\cos\theta}-\cos\theta\right)d\theta$

$= \pi\int_0^{\frac{1}{4}\pi}(\sec\theta-\cos\theta)\,d\theta$

(c) $\pi\int_0^{\frac{1}{4}\pi}(\sec\theta-\cos\theta)\,d\theta$

$= \pi\left[\ln(\sec\theta+\tan\theta)-\sin\theta\right]_0^{\frac{1}{4}\pi}$

$= \pi\left[\ln(\sqrt{2}+1)-\frac{1}{\sqrt{2}}\right]-\pi\ln 1$

$= \pi\left[\ln(1+\sqrt{2})-\frac{1}{2}\sqrt{2}\right]$

D Trapezium rule: percentage error (p 84)

D1 (a)

x	0.8	1.2	1.6	2
y	1.796	2.078	2.440	2.896

(b) $\frac{1}{2}(0.4)[(1.414+2.896)+2(1.579+1.796+2.078+2.440)] = 4.02$ (to 3 s.f.)

(c) Because the gradient of the curve is increasing, the area of each trapezium is greater than the area under the curve.

(d) Make h smaller and increase the number of trapeziums.

Exercise D (p 85)

1 3.75 (to 3 s.f)

2 (a) 2.73 (to 3 s.f.)

(b) 3.00 (to 3 s.f.)

(c) π

(d) (i) 13.0%

(ii) 4.6%

Increasing the ordinates from 3 to 5 more than halves the percentage error.

3 (a) 3.03 (to 3 s.f.)

(b) 2.97 (to 3 s.f.)

(c) $\int u\frac{dv}{dx}\,dx = uv-\int v\frac{du}{dx}\,dx$

$u = \ln x, \frac{du}{dx} = \frac{1}{x}, \frac{dv}{dx} = x, v = \frac{1}{2}x^2$

$\int_1^3 x\ln x\,dx = \left[\frac{1}{2}x^2\ln x\right]_1^3 - \int_1^3 \frac{1}{2}x^2\left(\frac{1}{x}\right)dx$

$= \frac{9}{2}\ln 3 - \frac{1}{2}\ln 0 - \frac{1}{2}\int_1^3 x\,dx$

$= \frac{9}{2}\ln 3 - \left[\frac{1}{4}x^2\right]_1^3$

$= \frac{9}{2}\ln 3 - \left(\frac{9}{4}-\frac{1}{4}\right)$

$= \frac{9}{2}\ln 3 - 2$

(d) (i) 3.1%

(ii) 0.8%

Using 5 ordinates gives a very accurate answer.

4 (a) 3.28 (to 3 s.f.)   **(b)** $\pi[3(\ln 3)^2 - 6\ln 3 + 4]$

(c) 1.4%

Mixed questions (p 86)

1 $\dfrac{dx}{d\theta} = 5\cos\theta,\ dx = 5\cos\theta\,d\theta$

$\displaystyle\int_0^5 y\,dx = \int_0^{\frac{1}{2}\pi} 5\cos\theta \times 5\cos\theta\,d\theta = 25\int_0^{\frac{1}{2}\pi}\cos^2\theta\,d\theta$

$= \dfrac{25}{2}\displaystyle\int_0^{\frac{1}{2}\pi}(\cos 2\theta + 1)\,d\theta = \dfrac{25}{2}\left[\tfrac{1}{2}\sin 2\theta + \theta\right]_0^{\frac{1}{2}\pi}$

$= \dfrac{25}{2}\left[\left(0 + \dfrac{\pi}{2}\right) - 0\right] = \dfrac{25\pi}{4}$

2 $\dfrac{\pi}{36}(23e^{24} - 5e^6)$

3 (a) 0.818 (to 3 s.f.)

(b) $x = 3\tan\theta,\ \dfrac{dx}{d\theta} = 3\sec^2\theta,\ dx = 3\sec^2\theta\,d\theta$

$9 + x^2 = 9 + 9\tan^2\theta = 9(1 + \tan^2\theta) = 9\sec^2\theta$

$\displaystyle\int_0^3 \pi y^2\,dx = \pi\int_0^3 \dfrac{1}{9 + x^2}\,dx$

$= \pi\displaystyle\int_0^{\frac{1}{4}\pi}\dfrac{1}{9\sec^2\theta}\times 3\sec^2\theta\,d\theta = \tfrac{1}{3}\pi\int_0^{\frac{1}{4}\pi}1\,d\theta$

$= \tfrac{1}{3}\pi[\theta]_0^{\frac{1}{4}\pi} = \dfrac{\pi^2}{12}$

(c) 0.6%

4 $x = \theta + \tan\theta,\ \dfrac{dx}{d\theta} = 1 + \sec^2\theta,\ dx = (1 + \sec^2\theta)\,d\theta$

$\displaystyle\int_0^{\frac{1}{4}\pi}\pi y^2\,dx = \pi\int_0^{\frac{1}{4}\pi}\cos^2\theta(1 + \sec^2\theta)\,d\theta$

$= \pi\displaystyle\int_0^{\frac{1}{4}\pi}(\cos^2\theta + 1)\,d\theta = \pi\int_0^{\frac{1}{4}\pi}\left(\tfrac{1}{2}(\cos 2\theta + 1) + 1\right)d\theta$

$= \tfrac{1}{2}\pi\displaystyle\int_0^{\frac{1}{4}\pi}(\cos 2\theta + 3)\,d\theta = \tfrac{1}{2}\pi\left[\tfrac{1}{2}\sin 2\theta + 3\theta\right]_0^{\frac{1}{4}\pi}$

$= \tfrac{1}{2}\pi\left[\tfrac{1}{2} + \tfrac{3}{4}\pi\right] = \dfrac{\pi}{4} + \dfrac{3\pi^2}{8}$

5 (a) $\displaystyle\int u\dfrac{dv}{dx}\,dx = uv - \int v\dfrac{du}{dx}\,dx$

$u = x^2,\ \dfrac{du}{dx} = 2x,\ \dfrac{dv}{dx} = e^{-x},\ v = -e^{-x}$

$\displaystyle\int_0^1 x^2 e^{-x}\,dx = \left[x^2(-e^{-x})\right]_0^1 - \int_0^1 (-e^{-x})2x\,dx$

$= -e^{-1} + 2\displaystyle\int_0^1 xe^{-x}\,dx$

$u = x,\ \dfrac{du}{dx} = 1,\ \dfrac{dv}{dx} = e^{-x},\ v = -e^{-x}$

$2\displaystyle\int_0^1 xe^{-x}\,dx = 2\left[x(-e^{-x})\right]_0^1 - 2\int_0^1 (-e^{-x})\,dx$

$= -2e^{-1} - 2\left[e^{-x}\right]_0^1 = -2e^{-1} - 2e^{-1} + 2e^0 = -4e^{-1} + 2$

$\displaystyle\int_0^1 x^2 e^{-x}\,dx = -e^{-1} + (-4e^{-1} + 2) = 2 - 5e^{-1}$

(b) $5\pi(1 - 2e^{-1})$

6 (a) $\tfrac{1}{12}(9 - 3\sqrt{2} + \sqrt{6})$ **(b)** $2 - \sqrt{2}$

(c) 2.5%

Test yourself (p 87)

1 3π

2 $\tfrac{1}{4}\pi(e^2 + 1)$

3 (a) 0.635 (to 3 s.f.)   **(b)** $2(1 - \ln 2)$

(c) 3.4%

4 (a) $\displaystyle\int u\dfrac{dv}{dx}\,dx = uv - \int v\dfrac{du}{dx}\,dx$

$u = x,\ \dfrac{du}{dx} = 1,\ \dfrac{dv}{dx} = \sec^2 x,\ v = \tan x$

$\displaystyle\int_0^{\frac{\pi}{4}} x\sec^2 x\,dx = \left[x\tan x\right]_0^{\frac{\pi}{4}} - \int_0^{\frac{\pi}{4}}\tan x\,dx$

$= \tfrac{1}{4}\pi - \left[\ln\sec x\right]_0^{\frac{\pi}{4}}$

$= \tfrac{1}{4}\pi - \left[\ln\sqrt{2} - \ln 1\right]$

$= \tfrac{1}{4}\pi - \tfrac{1}{2}\ln 2$

(b) $\tfrac{1}{4}\pi^2 - \tfrac{1}{2}\pi\ln 2$

(c) $\sqrt{\dfrac{2}{\pi}} + \sqrt{\dfrac{\pi}{2}}$

7 Integration 3

A Using partial fractions (p 88)

A1 (a) $\dfrac{2}{x-2}+\dfrac{3}{x+1}$ (b) $\dfrac{3}{x-2}-\dfrac{4}{2x+1}$

(c) $\dfrac{7}{4(x-3)}-\dfrac{7}{4(x+1)}-\dfrac{2}{(x+1)^2}$

A2 (a) $1+\dfrac{4}{x+1}-\dfrac{7}{x+2}$ (b) $1+\dfrac{4}{5(x-2)}-\dfrac{9}{5(x+3)}$

(c) $-2+\dfrac{5}{3(2-x)}-\dfrac{1}{3(x+1)}$

A3 $2\ln|x-2|+3\ln|x+1|+c$

A4 $3\ln|x-2|-2\ln|2x+1|+c$

A5 (a) $-\dfrac{3}{x}+\dfrac{4}{x+2}$

(b) $-3\ln|x|+4\ln|x+2|+c$

A6 (a) $x+4\ln|x+1|-7\ln|x+2|+c$

(b) $x+\frac{4}{5}\ln|x-2|-\frac{9}{5}\ln|x+3|+c$

(c) $-2x-\frac{5}{3}\ln|2-x|-\frac{1}{3}\ln|x+1|+c$

A7 $2\ln|x-1|+4\ln|x+1|+\dfrac{3}{x+1}+c$

A8 $\frac{7}{4}\ln|x-3|-\frac{7}{4}\ln|x+1|+\dfrac{2}{x+1}+c$

Exercise A (p 92)

1 (a) $-\ln|x|+\ln|x-1|+c$

(b) $\ln|x-1|-\ln|x+2|+c$

(c) $\frac{1}{4}\ln|x-1|+\frac{3}{4}\ln|x+3|+c$

(d) $\frac{6}{5}\ln|x-1|-\frac{1}{5}\ln|x+4|+c$

(e) $\frac{3}{2}\ln|2x-1|-\ln|x+2|+c$

(f) $-2\ln|1-x|-\frac{1}{2}\ln|2x+1|+c$

2 (a) $x+\ln|x+1|-4\ln|x+2|+c$

(b) $x-\ln|x|+4\ln|x-3|+c$

(c) $x-\frac{1}{3}\ln|x+2|-\frac{2}{3}\ln|x-1|+c$

3 (a) $\frac{2}{9}\ln|x-2|-\frac{2}{9}\ln|x+1|-\dfrac{1}{3(x+1)}+c$

(b) $\frac{1}{2}\ln|x+1|-\frac{1}{2}\ln|x-1|-\dfrac{2}{x-1}+c$

(c) $\ln|2x-1|-\ln|x-1|-\dfrac{1}{x-1}+c$

(d) $2\ln|x|+\dfrac{1}{x}-2\ln|x+1|+c$

(e) $\ln|x+2|-\dfrac{1}{x+2}-\ln|2x+3|+c$

(f) $-\frac{1}{16}\ln|x|+\frac{1}{16}\ln|x-4|-\dfrac{3}{4(x-4)}+c$

B Definite integrals

Exercise B (p 93)

1 (a) $\ln\frac{4}{3}$

(b) $\frac{1}{3}\ln\frac{16}{7}$

(c) $-\frac{7}{5}\ln 4$

(d) $\frac{1}{4}\ln 2+\frac{3}{4}\ln\frac{6}{5}$

(e) $\frac{5}{3}\ln 2-\frac{7}{6}\ln\frac{7}{5}$

(f) $-\ln 2-\frac{3}{2}\ln 3$

2 (a) $1+7\ln 2-4\ln 3=1+\ln\frac{128}{81}$

(b) $3+\frac{5}{3}\ln 4-\frac{4}{3}\ln 7$

(c) $2+9\ln 2-\frac{15}{2}\ln 3$

(d) $2-\frac{25}{8}\ln\frac{7}{5}-\frac{9}{8}\ln 3$

(e) $-2+3\ln 5-2\ln 3=-2+\ln\frac{125}{9}$

(f) $1-\ln 3$

3 (a) $\frac{1}{2}-\frac{1}{8}\ln 3$

(b) $\frac{4}{9}\ln 4-\frac{2}{9}\ln 7+\frac{1}{28}$

(c) $\frac{4}{3}-\ln 3+\frac{1}{2}\ln 5$

(d) $4\ln 2-2\ln 3-\frac{1}{2}=\ln\frac{16}{9}-\frac{1}{2}$

(e) $3\ln 5-3\ln 3-\frac{2}{3}=\ln\frac{125}{27}-\frac{2}{3}$

(f) $2\ln 5-3\ln 3+\frac{2}{15}=\ln\frac{25}{27}+\frac{2}{15}$

4 Given that $\dfrac{4}{(1+3x)(1-x)}\equiv\dfrac{A}{1+3x}+\dfrac{B}{1-x}$

we have $A(1-x)+B(1+3x)\equiv 4$

Putting $x=1$ gives $4B=4\implies B=1$ and
putting $x=-\frac{1}{3}$ gives $\frac{4}{3}A=4\implies A=3$

So $\displaystyle\int_0^{\frac{1}{3}}\dfrac{4}{(1+3x)(1-x)}\,\mathrm{d}x=\int_0^{\frac{1}{3}}\left(\dfrac{3}{1+3x}+\dfrac{1}{1-x}\right)\mathrm{d}x$

$=\left[\ln(1+3x)-\ln(1-x)\right]_0^{\frac{1}{3}}$

$=\left(\ln 2-\ln\frac{2}{3}\right)-(\ln 1-\ln 1)$

$=\ln 2-\ln\frac{2}{3}$

$=\ln 2-(\ln 2-\ln 3)$

$=\ln 3\quad$ as required.

5 Given that $\dfrac{9}{(1+x)(2-x)^2} \equiv \dfrac{A}{(2-x)^2} + \dfrac{B}{2-x} + \dfrac{C}{1+x}$

we have $A(1+x) + B(2-x)(1+x) + C(2-x)^2 \equiv 9$

Putting $x = -1$ gives $9C = 9 \Rightarrow C = 1$

Putting $x = 2$ gives $3A = 9 \Rightarrow A = 3$

Putting $x = 0$ gives $A + 2B + 4C = 9$

$\Rightarrow \quad 3 + 2B + 4 = 9$

$\Rightarrow \quad\quad\quad B = 1$

So $\displaystyle\int_{-\frac{1}{2}}^{\frac{1}{2}} \dfrac{9}{(1+x)(2-x)^2}\,dx$

$= \displaystyle\int_{-\frac{1}{2}}^{\frac{1}{2}} \left(\dfrac{3}{(2-x)^2} + \dfrac{1}{2-x} + \dfrac{1}{1+x} \right) dx$

$= \left[\dfrac{3}{2-x} - \ln(2-x) + \ln(1+x) \right]_{-\frac{1}{2}}^{\frac{1}{2}}$

$= \left(2 - \ln\left(\tfrac{3}{2}\right) + \ln\left(\tfrac{3}{2}\right) \right) - \left(\tfrac{6}{5} - \ln\left(\tfrac{5}{2}\right) + \ln\left(\tfrac{1}{2}\right) \right)$

$= 2 - \tfrac{6}{5} + \ln\left(\tfrac{5}{2}\right) - \ln\left(\tfrac{1}{2}\right)$

$= \tfrac{4}{5} + \ln\left(\dfrac{\frac{5}{2}}{\frac{1}{2}}\right)$

$= \tfrac{4}{5} + \ln 5 \quad$ as required

6 $\tfrac{1}{2} + \ln\left(\tfrac{9}{4}\right)$

Mixed questions (p 93)

1 (a) $\dfrac{1}{x} + \dfrac{2}{x+3}$

(b) $\displaystyle\int_1^4 \dfrac{3(x+1)}{x(x+3)}\,dx = \int_1^4 \left(\dfrac{1}{x} + \dfrac{2}{x+3} \right) dx$

$= \left[\ln x + 2\ln(x+3) \right]_1^4$

$= (\ln 4 + 2\ln 7) - (\ln 1 + 2\ln 4)$

$= \ln 4 + 2\ln 7 - 2\ln 4$

$= 2\ln 7 - \ln 4$

2 (a) $A = \tfrac{1}{2}, B = -\tfrac{1}{2}$ **(b)** $\ln\left(\dfrac{3}{\sqrt{5}}\right)$

3 (a) $A = 1, B = 1, C = -3$

(b) $x + \ln|x+1| - \tfrac{3}{2}\ln|2x-1| + c$

(c) $\displaystyle\int_1^2 \dfrac{2x^2-5}{(x+1)(2x-1)}\,dx = \left[x + \ln(x+1) - \tfrac{3}{2}\ln(2x-1) \right]_1^2$

$= \left(2 + \ln 3 - \tfrac{3}{2}\ln 3 \right) - \left(1 + \ln 2 - \tfrac{3}{2}\ln 1 \right)$

$= 2 - 1 - \tfrac{1}{2}\ln 3 - \ln 2$

$= 1 - \left(\ln\sqrt{3} + \ln 2 \right)$

$= 1 - \ln\left(2\sqrt{3}\right) \quad$ as required

4 (a) $\dfrac{2}{(x+1)^2} - \dfrac{1}{x+1} + \dfrac{2}{2x-3}$

(b) $\tfrac{1}{6} + 2\ln\tfrac{3}{2}$

5 $4 + \ln\tfrac{5}{7}$

6 $-\tfrac{13}{8}\ln 5 + 2\ln 6$

7 (a) $3\ln|x-1| - 3\ln|x+1| + c$ or $3\ln\left|\dfrac{x-1}{x+1}\right| + c$

(b) $\ln|x^2 - 1| + c$ or $\ln|x+1| + \ln|x-1| + c$

(c) $x - \tfrac{1}{5}\ln|x| + \tfrac{26}{5}\ln|x-5| + c$

(d) $\ln|x^2 - 5x| + c$ or $\ln|x| + \ln|x-5| + c$

(e) $\tfrac{1}{2}\ln|x^2 + 1| + c$

(f) $6\arctan x + c$

(g) $x - \arctan x + c$

(h) $-\dfrac{\ln x}{x-1} + \ln|x-1| - \ln|x| + c$

Test yourself (p 95)

1 (a) $\tfrac{1}{2}\ln|x| - \tfrac{1}{6}\ln|3x+2| + c$

(b) $\tfrac{1}{3}x + \tfrac{1}{2}\ln|x| - \tfrac{13}{18}\ln|3x+2| + c$

(c) $-\tfrac{1}{4}\ln|x| - \dfrac{1}{2x} + \tfrac{1}{4}\ln|3x+2| + c$

2 (a) $\tfrac{1}{12}\ln 5 + \tfrac{1}{4}\ln 3 - \tfrac{1}{3}\ln 2$

(b) $\tfrac{1}{3} + \tfrac{1}{36}\ln 5 - \tfrac{1}{4}\ln 3 + \tfrac{2}{9}\ln 2$

(c) $\tfrac{1}{12} - \tfrac{1}{8}\ln 5 + \tfrac{1}{8}\ln 3$

3 (a) $A = -1, B = 1, C = 4$

(b) $-3 + 10\ln 2$

4 (a) $A = 16, B = -2, C = -1$

(b) $\tfrac{32}{15} - \ln 3$

5 Given that

$\dfrac{2x+1}{(1-2x)^2(x+1)} \equiv \dfrac{A}{(1-2x)^2} + \dfrac{B}{1-2x} + \dfrac{C}{x+1}$

we have $A(x+1) + B(1-2x)(x+1) + C(1-2x)^2$
$= 2x + 1$

Putting $x = -1$ gives $9C = -1 \Rightarrow C = -\tfrac{1}{9}$

Putting $x = \tfrac{1}{2}$ gives $\tfrac{3}{2}A = 2 \Rightarrow A = \tfrac{4}{3}$

Putting $x = 0$ gives $A + B + C = 1 \Rightarrow B = -\tfrac{2}{9}$

So $\displaystyle\int_0^{\frac{1}{3}} \frac{2x+1}{(1-2x)^2(x+1)}\,dx$

$= \displaystyle\int_0^{\frac{1}{3}} \left(\frac{4}{3(1-2x)^2} - \frac{2}{9(1-2x)} - \frac{1}{9(x+1)} \right) dx$

$= \dfrac{1}{9} \displaystyle\int_0^{\frac{1}{3}} \left(\frac{12}{(1-2x)^2} - \frac{2}{1-2x} - \frac{1}{x+1} \right) dx$

$= \dfrac{1}{9} \left[\dfrac{6}{1-2x} + \ln(1-2x) - \ln(x+1) \right]_0^{\frac{1}{3}}$

$= \dfrac{1}{9} \left(\left(18 + \ln\left(\tfrac{1}{3}\right) - \ln\left(\tfrac{4}{3}\right) \right) - \left(6 + \ln 1 - \ln 1 \right) \right)$

$= \dfrac{1}{9} \left(12 - \left(\ln\tfrac{4}{3} - \ln\tfrac{1}{3} \right) \right)$

$= \dfrac{1}{9} (12 - \ln 4)$

$= \dfrac{1}{9} (12 - 2\ln 2)$

$= \dfrac{4}{3} - \dfrac{2}{9} \ln 2$ as required

8 Differential equations

A Forming a differential equation (p 96)

A1 $\dfrac{dP}{dt} = 0.05P^2$

A2 $\dfrac{dT}{dt} = 2t + 0.01t^2$

A3 The differential equation, $\dfrac{dP}{dt} = 0.1Pt$, tells us that P is increasing at the rate $0.1Pt$. It should be $\dfrac{dP}{dt} = -0.1Pt$.

A4 $\dfrac{ds}{dt} = -0.02s$

A5 $\dfrac{dV}{dt} = 0.05t^2 - 4$

A6 $\dfrac{dA}{dt} = kA$

A7 $\dfrac{dh}{dt} = -k\sqrt{h}$

A8 $\dfrac{dA}{dt} = 2\pi r \dfrac{dr}{dt}$

A9 $\dfrac{dV}{dt} = 4\pi r^2 \dfrac{dr}{dt}$

A10 $\dfrac{ds}{dt} = \dfrac{1}{2} A^{-\frac{1}{2}} \dfrac{dA}{dt}$ or $\dfrac{ds}{dt} = \dfrac{1}{2\sqrt{A}} \dfrac{dA}{dt}$

A11 (a) $\dfrac{ds}{dt} = ks$

(b) $\dfrac{dA}{dt} = 2s \dfrac{ds}{dt}$

(c) From part (a) and (b) we have
$\dfrac{dA}{dt} = 2s \times ks = 2ks^2 = 2kA$ as $A = s^2$

Exercise A (p 99)

1 $\dfrac{dV}{dt} = 200 - 0.1V$; multiplying both sides by -10 gives $-10\dfrac{dV}{dt} = V - 2000$ as required.

2 (a) $I = \dfrac{k}{x^2}$

(b) $\dfrac{dI}{dt} = -\dfrac{2k}{x^3} \dfrac{dx}{dt}$

(c) The speed is $\dfrac{dx}{dt}$ and this is proportional to the distance x so there exists a constant b such that $\dfrac{dx}{dt} = bx$.

Hence $\dfrac{dI}{dt} = \dfrac{dI}{dx}\dfrac{dx}{dt}$

$$\Rightarrow \dfrac{dI}{dt} = -\dfrac{2k}{x^3} \times bx$$

$$= -2b \times \dfrac{k}{x^2}$$

$$= -2bI$$

$$= -aI$$

where $a = 2b$ and is a constant.

3 (a) The diagram shows the cross-section.

The larger triangle is equilateral so $\theta = 30°$. Hence the base of the smaller triangle is $2h\tan 30° = \dfrac{2h}{\sqrt{3}}$ and so the volume of water in the tank is $\tfrac{1}{2} \times \dfrac{2h}{\sqrt{3}} \times h \times l$, where l is the length of the tank. So $V = \dfrac{l}{\sqrt{3}}h^2$, where $\dfrac{l}{\sqrt{3}}$ is a constant, and so $V \propto h^2$ as required.

(b) Since the water flows in at a constant rate, there exists a constant k such that $\dfrac{dV}{dt} = k$.

Also $V = \dfrac{l}{\sqrt{3}}h^2$ gives

$$\dfrac{dV}{dh} = \dfrac{2l}{\sqrt{3}}h$$

So $\dfrac{dV}{dt} = \dfrac{dV}{dh}\dfrac{dh}{dt}$

$$\Rightarrow k = \dfrac{2l}{\sqrt{3}}h\dfrac{dh}{dt}$$

$$\Rightarrow \dfrac{dh}{dt} = \dfrac{\sqrt{3}k}{2l} \times \dfrac{1}{h}$$

As $\dfrac{\sqrt{3}k}{2l}$ is a constant, $\dfrac{dh}{dt}$ is inversely proportional to h.

4 (a) $\dfrac{dV}{dt} = h^3\dfrac{dh}{dt}$

(b) $\dfrac{dV}{dt} \propto \sqrt{V}$ so there exists a constant k such that $\dfrac{dV}{dt} = k\sqrt{V}$. The result from (a) gives

$$h^3\dfrac{dh}{dt} = k\sqrt{V}$$

Substituting $V = 0.25h^4 = \dfrac{h^4}{4}$ gives

$$h^3\dfrac{dh}{dt} = k\sqrt{\dfrac{h^4}{4}}$$

$$\Rightarrow h^3\dfrac{dh}{dt} = \dfrac{k}{2}h^2$$

$$\Rightarrow \dfrac{dh}{dt} = \dfrac{k}{2} \times \dfrac{1}{h}$$

As $\dfrac{k}{2}$ is a constant, $\dfrac{dh}{dt}$ is inversely proportional to h.

5 $E \propto v^2$ so there exists a constant k such that $E = kv^2$. The speed is increasing at a constant rate so there exists a constant c such that $\dfrac{dv}{dt} = c$.

Hence $\dfrac{dE}{dt} = \dfrac{dE}{dv} \times \dfrac{dv}{dt}$

$$\Rightarrow \dfrac{dE}{dt} = 2kv \times c$$

$$= 2ck\sqrt{\dfrac{E}{k}}$$

$$= 2c\sqrt{k}\sqrt{E}$$

As $2c\sqrt{k}$ is a constant, $\dfrac{dE}{dt} \propto \sqrt{E}$.

B Solving by separating variables (p 100)

B1 $\dfrac{dP}{dt} = 0.1Ae^{0.1t} = 0.1P$

B2 (a) (i) $P\,dP = 5\,dt$

(ii) $\int P\,dP = \int 5\,dt \Rightarrow \tfrac{1}{2}P^2 = 5t + c$

$P^2 = 10t + 2c$

If $A = 2c$ then $P^2 = 10t + A$

(b) $\dfrac{dP}{dt} = \tfrac{1}{2} \times 10(10t + A)^{-\frac{1}{2}} = \dfrac{5}{\sqrt{10t + A}} = \dfrac{5}{P}$

(c) Substituting $P = 6$ and $t = 0$ in $P^2 = 10t + A$ gives $36 = 10 \times 0 + A \Rightarrow A = 36$

(d) $P^2 = 10 \times 10.8 + 36 = 108 + 36 = 144$

$P = 12$ $(P > 0)$

B3 (a) (i) $\frac{1}{P^2}\,dP = 0.01\,dt$

(ii) $\int \frac{1}{P^2}\,dP = \int 0.01\,dt \Rightarrow -\frac{1}{P} = 0.01t + c$

Rearranging gives $P = -\dfrac{1}{0.01t + c}$.

(b) Substituting $P = 10$ and $t = 0$ in

$P = -\dfrac{1}{0.01t + c}$ gives $10 = -\dfrac{1}{0.01\times 0 + c} = -\dfrac{1}{c}$

so $c = -0.1$. Hence

$P = -\dfrac{1}{0.01t - 0.1} \Rightarrow P = \dfrac{100}{10 - t}$

(c) $P = \dfrac{100}{10 - t}$

As t gets closer to 10, $10 - t$ gets closer to 0 and P gets larger.

B4 (a) You cannot integrate y^2 with respect to x. The student should have rearranged the equation so there in only one variable on each side.

(b) $dy = 3y^2\,dx$

$\int \frac{1}{y^2}\,dy = \int 3\,dx$

$-\dfrac{1}{y} = 3x + c$

Rearranging this gives $y = -\dfrac{1}{3x + c}$.

Exercise B (p 102)

1 (a) Separating the variables, $y\,dy = x^3\,dx$

Integrating both sides, $\int y\,dy = \int x^3\,dx$

$\Rightarrow \frac{1}{2}y^2 = \frac{1}{4}x^4 + c \Rightarrow y^2 = \frac{1}{2}x^4 + A$

(where $A = 2c$)

(b) $y^2 = \frac{1}{2}x^4 + 16$

2 (a) Separating the variables, $y\,dy = (1 + x)\,dx$

Integrating both sides, $\int y\,dy = \int (1 + x)\,dx$

$\Rightarrow \frac{1}{2}y^2 = x + \frac{1}{2}x^2 + c \Rightarrow y^2 = x^2 + 2x + A$

(where $A = 2c$)

(b) 6

3 (a) Separating the variables, $\dfrac{1}{y}\,dy = x\,dx$

Integrating both sides, $\int \frac{1}{y}\,dy = \int x\,dx$

$\Rightarrow \log y = \frac{1}{2}x^2 + c \Rightarrow y = e^{\frac{1}{2}x^2 + c} = Ae^{\frac{1}{2}x^2}$

(where $A = e^c$)

(b) 6

4 $y = 6e^x - 1$

5 (a) Separating the variables, $\dfrac{dy}{\sqrt{y}} = 2x\,dx$

Integrating both sides, $\int \dfrac{dy}{\sqrt{y}} = \int 2x\,dx$

$\Rightarrow 2\sqrt{y} = x^2 + c$

$\Rightarrow \sqrt{y} = \frac{1}{2}x^2 + A$ (where $A = \frac{1}{2}c$)

(b) 2   **(c)** 2.22

6 (a) $\dfrac{dP}{dt} = -kP$

(b) Separating the variables, $\dfrac{dP}{P} = -k\,dt$

Integrating both sides, $\int \dfrac{dP}{P} = -\int k\,dt$

$\Rightarrow \ln P = -kt + c$

$\Rightarrow P = e^{-kt + c} = Ae^{-kt}$ (where $A = e^c$)

(c) 20 000   **(d)** 0.0693

(e) 43.2 (to 3 s.f.)

7 (a) $y = Ae^{-\frac{1}{x}}$ **(b)** $y = e^2e^{-\frac{1}{x}} = e^{2 - \frac{1}{x}}$

8 $y = \dfrac{2}{2 - x^2}$

9 $y = \dfrac{1}{2 - \ln x}$

10 (a) $\frac{1}{2}\ln(x^2 + 1) + c$

(b) (i) Separating the variables, $\dfrac{dy}{y} = \dfrac{x\,dx}{x^2 + 1}$

$\Rightarrow \ln y = \frac{1}{2}\ln(x^2 + 1) + c$

$\ln y = \ln\sqrt{x^2 + 1} + \ln A = \ln\left(A\sqrt{x^2 + 1}\right)$

(where $\ln A = c$)

$\Rightarrow y = A\sqrt{x^2 + 1}$

(ii) $\dfrac{10}{\sqrt{2}}$, or $5\sqrt{2}$

(c) (i) $\dfrac{1}{y} = c - \frac{1}{2}\ln(x^2 + 1)$ $\left(y = \dfrac{1}{c - \frac{1}{2}\ln(x^2 + 1)}\right)$

(ii) $1 = c - \frac{1}{2}\ln 1 \Rightarrow c = 1$

$\Rightarrow \dfrac{1}{y} = 1 - \frac{1}{2}\ln(x^2 + 1)$

$\Rightarrow y = \dfrac{1}{1 - \frac{1}{2}\ln(x^2 + 1)}$

11 (a) $\dfrac{dT}{dt} = -k(T - A)$ $(T - A > 0, k > 0)$

  (b) Separating the variables, $\dfrac{dT}{T - A} = -k\,dt$

  $\Rightarrow \ln(T - A) = -kt + c$

  $T - A = e^{-kt + c} = Be^{-kt}$ (where $B = e^c$)

  $T = A + Be^{-kt}$

C Exponential growth and decay (p 104)

C1 (a) 4.05   **(b)** 6.93   **(c)** −2.88

C2 (a) (i) 1.02   **(ii)** 1.83   **(iii)** −0.365

  (b) P gets closer and closer to 0.

C3 (a) 50

  (b) As $t \to \infty$, $e^{-2t} \to 0$, so the limiting value is 40.

  (c)
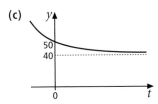

C4 (a) As $t \to \infty$, $e^{-bt} \to 0$ so $T \to c + a(0) = c$
  This will be the temperature of the surrounding atmosphere, so $c = 20$.

  (b) When $t = 0$, $80 = 20 + a$ so $a = 60$

  (c) When $t = 10$, $35 = 20 + 60e^{-10b}$

  $60e^{-10b} = 15$

  $e^{-10b} = 0.25$

  $-10b = \ln(0.25)$

  $b = 0.139$ (to 3 s.f.)

  (d) $\dfrac{dT}{dt} = 60 \times (-0.139e^{-0.139t}) = -8.34e^{-1.39}$

  $= -2.08$ degrees per minute (to 3 s.f.)

Exercise C (p 106)

1 (a) 5000   **(b)** 11 128   **(c)** 445

2 (a) 40   **(b)** 50

3 (a) 800

  (b) $500 = 800e^{-5k} \Rightarrow e^{-5k} = 0.625$

  $k = -\tfrac{1}{5}\ln 0.625 = 0.0940$ (to 3 s.f.)

  (c) (i) 549   **(ii)** 7.37

4 (a) 185   **(b)** $T \to 200$ **(c)** 0.0615

5 $N = \tfrac{1}{2}N_0 = N_0e^{-0.04t}$

  $e^{-0.04t} = 0.5$

  $-0.04t = \ln(0.5) \Rightarrow t = \dfrac{\ln(0.5)}{-0.04} = 17.3$

D Further exponential functions (p 106)

D1 (a) $5^x = (e^{\ln 5})^x = e^{(\ln 5)x} = e^{1.61x}$

  (b) $e^{(\ln 8)x} = e^{2.08x}$

  (c) $e^{0.693x}$ **(d)** $e^{(\ln a)x}$

D2 $\dfrac{d(a^x)}{dx} = \dfrac{d}{dx}(e^{(\ln a)x}) = (\ln a)e^{(\ln a)x} = (\ln a)a^x$

D3 (a) £20 000

  (b) £8874

  (c) $t = \dfrac{\ln 0.4}{\ln 0.85} = 5.64$

  (d) $20\,000 \times \ln 0.85 \times 0.85^5 = -1440$ (to 3 s.f.)

D4 (a) 1.2   **(b)** 8000×1.2^t

D5 (a) 0.8   **(b)** 8000×0.8^t

Exercise D (p 108)

1 (a) 5.69 years   **(b)** 11 012 per year

2 (a) 3.11 years   **(b)** £2856 per year

3 $\dfrac{dy}{dx} = qa^x \ln a = (y - p)\ln a$

Mixed questions (p 108)

1 (a) $\dfrac{dN}{dt} = kN(20\,000 - N)$

  (b) The increase in sales stops when $\dfrac{dN}{dt} = 0$, i.e. when $N = 20\,000$.

  When $N > 20\,000$, $\dfrac{dN}{dt} < 0$ and total sales would then decrease. As this cannot happen in practice, 20 000 is the limiting value for N in the model.

2 (a) $\dfrac{dr}{dt} \propto r$ so there exists a constant k such that

$\dfrac{dr}{dt} = kr$. From $V = \frac{4}{3}\pi r^3$ we have $\dfrac{dV}{dr} = 4\pi r^2$.

$$\dfrac{dV}{dt} = \dfrac{dV}{dr}\dfrac{dr}{dt}$$

$$\Rightarrow \quad \dfrac{dV}{dt} = 4\pi r^2 \times kr$$

$$= 4k\pi r^3$$

$$= 3k \times \tfrac{4}{3}\pi r^3$$

$$= 3kV$$

$$= aV$$

where $a = 3k$ so a is a constant

(b) $\dfrac{dV}{dt} \propto -V$ so these exists a positive constant k

such that $\dfrac{dV}{dt} = -kV$.

$$\dfrac{dV}{dt} = \dfrac{dV}{dr}\dfrac{dr}{dt}$$

$$\Rightarrow \quad -kV = 4\pi r^2 \dfrac{dr}{dt}$$

$$\Rightarrow -k \times \tfrac{4}{3}\pi r^3 = 4\pi r^2 \dfrac{dr}{dt}$$

$$\Rightarrow \quad -\tfrac{1}{3}kr = \dfrac{dr}{dt}$$

So $\dfrac{dr}{dt} = -br$ where $b = \frac{1}{3}k$, so b is a positive constant since k is a positive constant.

3 $y = x$

4 (a) $\dfrac{dh}{dt} = \dfrac{k}{h^2}$

(b) Separating the variables, $h^2\, dh = k\, dt$

Integrating both sides,

$\int h^2\, dh = \int k\, dt \Rightarrow \frac{1}{3}h^3 = kt + c$

$h^3 = 3kt + 3c$

so $h^3 = At + B$ $(A = 3k, B = 3c)$

(c) $A = 61, B = 64$

5 (a) (i) $\dfrac{dS}{dt} = 8\pi r \dfrac{dr}{dt}$

(ii) $\dfrac{dr}{dt} = \frac{1}{4}r$ so

$$\dfrac{dS}{dt} = 8\pi r \times \tfrac{1}{4}r$$

$$= 2\pi r^2$$

$$= \tfrac{1}{2}S$$

(b) $S = 5e^{\frac{1}{2}t}$

6 (a) $A = 3, B = -2$

(b) $y = 3\ln(x-2) - 2\ln(x-1) + \ln 4$ or

$$y = \ln\left(\dfrac{4(x-2)^3}{(x-1)^2}\right)$$

7 (a) Let r be the radius of the base of the tank. Then, at t seconds, the volume of water in the tank is $V = \pi r^2 h$. So $\dfrac{dV}{dh} = \pi r^2 = \dfrac{V}{h}$. The rate at which water flows in is inversely proportional to V so there exists a constant a such that $\dfrac{dV}{dt} = \dfrac{a}{V}$.

$$\dfrac{dV}{dt} = \dfrac{dV}{dh}\dfrac{dh}{dt}$$

$$\Rightarrow \quad \dfrac{a}{V} = \dfrac{V}{h}\dfrac{dh}{dt}$$

$$\Rightarrow \quad \dfrac{dh}{dt} = \dfrac{ah}{V^2}$$

$$= \dfrac{ah}{\left(\pi r^2 h\right)^2}$$

$$= \dfrac{ah}{\left(\pi r^2\right)^2 h^2}$$

$$= \dfrac{a}{\left(\pi r^2\right)^2} \times \dfrac{1}{h}$$

$$= \dfrac{k}{h}$$

where $k = \dfrac{a}{\left(\pi r^2\right)^2}$ and so k is a constant.

(b) $\dfrac{dh}{dt} = \dfrac{k}{h}$ gives $\int h\, dh = \int k\, dt$

$\Rightarrow \frac{1}{2}h^2 = kt + c$ where c is a constant

$\Rightarrow h^2 = 2kt + 2c$

$\Rightarrow h = \sqrt{2kt + 2c}$ as $h > 0$

$\Rightarrow h = \sqrt{At + B}$ where $A = 2k$ and $B = 2c$ and so A and B are constants.

(c) $A = \frac{21}{5}$ and $B = 4$

8 $y = \left(\frac{1}{3}x^{\frac{3}{2}} + 1\right)^2$

9 $y = e^{\frac{1}{2}x^2 + x} - 1$

10 (a) $\frac{1}{2}\ln(1 + y^2) + c$ **(b)** $y^2 = e^{2x} - 1$

11 $\dfrac{dy}{dx} = \cos^2 x \cos^2 y$

$\Rightarrow \displaystyle\int \dfrac{1}{\cos^2 y}\,dy = \int \cos^2 x\,dx$

$\Rightarrow \displaystyle\int \sec^2 y\,dy = \tfrac{1}{2}\int (\cos(2x) + 1)\,dx$

$\Rightarrow \tan y = \tfrac{1}{4}\sin(2x) + \tfrac{1}{2}x + c$ where c is a constant

$\Rightarrow y = \arctan\left(\tfrac{1}{4}\sin(2x) + \tfrac{1}{2}x + c\right)$ where c is a constant

12 (a) $f \propto \dfrac{1}{L}$ so there exists a constant a such that

$f = \dfrac{a}{L}$. Since $\dfrac{dL}{dt} \propto \sqrt{L}$, there exists a constant b

such that $\dfrac{dL}{dt} = b\sqrt{L}$.

$f = \dfrac{a}{L} \;\Rightarrow\; \dfrac{df}{dL} = -\dfrac{a}{L^2}$

$\dfrac{df}{dt} = \dfrac{df}{dL}\dfrac{dL}{dt}$

$\Rightarrow \dfrac{df}{dt} = -\dfrac{a}{L^2} \times b\sqrt{L}$

$= -abL^{-\frac{3}{2}}$

Now $f = \dfrac{a}{L} \;\Rightarrow\; L = \dfrac{a}{f} = af^{-1}$

So $\dfrac{df}{dt} = -ab(af^{-1})^{-\frac{3}{2}}$

$= -a^{-\frac{1}{2}}bf^{\frac{3}{2}}$

$= -kf^{\frac{3}{2}}$

where $k = a^{-\frac{1}{2}}b$ and so is a constant

(b) $\dfrac{df}{dt} = -kf^{\frac{3}{2}}$

$\Rightarrow \displaystyle\int -f^{-\frac{3}{2}}\,df = \int k\,dt$

$\Rightarrow 2f^{-\frac{1}{2}} = kt + a$ where a is a constant

$\Rightarrow \dfrac{2}{\sqrt{f}} = kt + a$

$\Rightarrow \sqrt{f} = \dfrac{2}{kt + a}$

$\Rightarrow f = \left(\dfrac{2}{kt + a}\right)^2 = \dfrac{4}{(kt + a)^2}$

$= \dfrac{4}{(kt + c)^2}$

where $c = a$ and so c is a constant

13 (a) Separating the variables and integrating,

$-\ln(50 - r) = \tfrac{1}{2}t + c$

$\ln(50 - r) = -\tfrac{1}{2}t - c$

$50 - r = e^{-\frac{1}{2}t - c}$

$r = 50 - e^{-\frac{1}{2}t - c} = 50 - Ae^{-\frac{1}{2}t}$ (where $A = e^{-c}$)

(b) $r \to 50$ **(c)** 30   **(d)** 2.20

14 (a) $\dfrac{1}{y^2 - 1} = \dfrac{1}{2(y-1)} - \dfrac{1}{2(y+1)}$

(b) $\dfrac{dy}{dx} = y^2 - 1$

$\Rightarrow \displaystyle\int \dfrac{1}{y^2 - 1}\,dy = \int 1\,dx$

$\Rightarrow \displaystyle\int \left(\dfrac{1}{2(y-1)} - \dfrac{1}{2(y+1)}\right)dy = \int 1\,dx$

$\Rightarrow \tfrac{1}{2}\ln(y-1) - \tfrac{1}{2}\ln(y+1) = x + c$

where c is a constant

$\Rightarrow \ln\left(\dfrac{y-1}{y+1}\right) = 2x + 2c$

$\Rightarrow \dfrac{y-1}{y+1} = e^{2x+2c} = e^{2c}e^{2x}$

$= Ae^{2x}$ where $A = e^{2c}$

and so A is a constant

From this we have $(y + 1)Ae^{2x} = y - 1$

$\Rightarrow yAe^{2x} + Ae^{2x} = y - 1$

$\Rightarrow Ae^{2x} + 1 = y(1 - Ae^{2x})$

$\Rightarrow y = \dfrac{1 + Ae^{2x}}{1 - Ae^{2x}}$

where A is a constant

Test yourself (p 111)

1 (a) $y = 3 + Ae^{\frac{1}{2}x}$ **(b)** $y = 3 - e^{\frac{1}{2}x}$

2 Separating the variables, $\dfrac{dy}{y+1} = x\,dx$

$\Rightarrow \ln(y+1) = \tfrac{1}{2}x^2 + c$

$\Rightarrow y + 1 = e^{\frac{1}{2}x^2 + c}$

$\Rightarrow y = e^{\frac{1}{2}x^2 + c} - 1 = Ae^{\frac{1}{2}x^2} - 1$ (where $A = e^c$)

3 (a) $\dfrac{13 - 2x}{(2x-3)(x+1)} \equiv \dfrac{4}{2x-3} - \dfrac{3}{x+1}$

(b) $y = \dfrac{108(2x-3)^2}{(x+1)^3}$

4 (a) $\dfrac{\mathrm{d}r}{\mathrm{d}t} \propto \dfrac{1}{r^2}$ so there is a constant k such that
$\dfrac{\mathrm{d}r}{\mathrm{d}t} = \dfrac{k}{r^2}$.

The stain is circular so $A = \pi r^2$ giving
$\dfrac{\mathrm{d}A}{\mathrm{d}r} = 2\pi r$

$$\dfrac{\mathrm{d}A}{\mathrm{d}t} = \dfrac{\mathrm{d}A}{\mathrm{d}r}\dfrac{\mathrm{d}r}{\mathrm{d}t}$$

$$\Rightarrow \quad \dfrac{\mathrm{d}A}{\mathrm{d}t} = 2\pi r \times \dfrac{k}{r^2}$$

$$= \dfrac{2\pi k}{r}$$

$$= \dfrac{2\pi k}{\sqrt{\dfrac{A}{\pi}}}$$

$$= 2\pi\sqrt{\pi}k \times \dfrac{1}{\sqrt{A}}$$

$$= 2\pi^{\frac{3}{2}}k \times \dfrac{1}{\sqrt{A}}$$

Now $2\pi^{\frac{3}{2}}k$ is a constant so $\dfrac{\mathrm{d}A}{\mathrm{d}t} \propto \dfrac{1}{\sqrt{A}}$ as required.

(b) 1.6 seconds

9 Vectors

A Vectors in two dimensions (p 112)

A1 No

A2 (a) 105°   **(b)** 55.9 km

A3 (a)

2c

(b)

$\frac{1}{2}$a

(c)

$-\frac{1}{3}$b

(d)

a   b   a + b

(e)

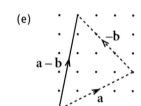

$-$b   a $-$ b   a

(f)

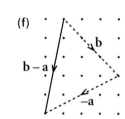

b   b $-$ a $-$a

(g)

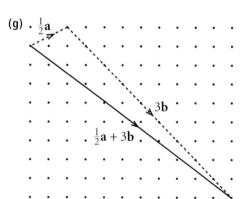

$\frac{1}{2}$a   3b $\frac{1}{2}$a + 3b

(h)

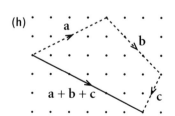

a   b   a + b + c   c

A4 Any two non-parallel vectors **x** and **y** can be used to form a parallelogram as follows.

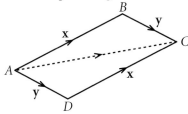

$\overrightarrow{AB} + \overrightarrow{BC} = \overrightarrow{AC}$ and $\overrightarrow{AD} + \overrightarrow{DC} = \overrightarrow{AC}$, showing that **x** + **y** = **y** + **x**.

If **x** and **y** are parallel then clearly it holds that **x** + **y** = **y** + **x**.

A5 (a) **a** + 5**b**   (b) 5**a** − 4**b**

A6 2**p** + 4**q** = 2(**p** + 2**q**) so the vectors **p** + 2**q** and 2**p** + 4**q** are parallel.

A7 B (6**x** − 3**y**), D (**y** − 2**x**) and E $(\mathbf{x} - \frac{1}{2}\mathbf{y})$

Exercise A (p 115)

1 (a) 3**x** + **y**   (b) 4**y** − **x**   (c) $\frac{1}{2}\mathbf{y}$

2 A (2**p** − 6**q**) and D $(\mathbf{q} - \frac{1}{3}\mathbf{p})$

3 (a) **a** + **b**   (b) **a** + **b** + **c**   (c) −**c**
   (d) 3**b** − **c**   (e) **a** − 2**b** + **c**

4 (a) (i) 2**p**   (ii) **p**   (iii) **q** − **p**   (iv) **q** − **p**
   (b) $\overrightarrow{AB} = \overrightarrow{DM} = \mathbf{p}$ and $\overrightarrow{AD} = \overrightarrow{BM} = \mathbf{q} - \mathbf{p}$
   Hence *ABMD* is a parallelogram.

5 (a) (i) $\overrightarrow{PR} = \overrightarrow{PQ} + \overrightarrow{QR} = \mathbf{a} + \mathbf{b}$ and
      $\overrightarrow{WX} = \frac{1}{2}\mathbf{a} + \frac{1}{2}\mathbf{b} = \frac{1}{2}(\mathbf{a} + \mathbf{b})$
      so *WX* is parallel to *PR*.
   (ii) $\overrightarrow{PR} = \overrightarrow{PS} + \overrightarrow{SR} = -\mathbf{d} - \mathbf{c}$ and
      $\overrightarrow{ZY} = -\frac{1}{2}\mathbf{d} - \frac{1}{2}\mathbf{c} = \frac{1}{2}(-\mathbf{d} - \mathbf{c})$
      so *ZY* is parallel to *PR*.
   (iii) Since *WX* and *ZY* are both parallel to *PR*, they must be parallel to each other.
   (b) $\overrightarrow{QS} = \overrightarrow{QR} + \overrightarrow{RS} = \mathbf{b} + \mathbf{c}$ and $\overrightarrow{XY} = \frac{1}{2}\mathbf{b} + \frac{1}{2}\mathbf{c}$
      $= \frac{1}{2}(\mathbf{b} + \mathbf{c})$ so *XY* is parallel to *QS*.
      $\overrightarrow{QS} = \overrightarrow{QP} + \overrightarrow{PS} = -\mathbf{a} - \mathbf{d}$ and $\overrightarrow{WZ} = -\frac{1}{2}\mathbf{a} - \frac{1}{2}\mathbf{d}$
      $= \frac{1}{2}(-\mathbf{a} - \mathbf{d})$ so *WZ* is parallel to *QS*.
      Since *XY* and *WZ* are both parallel to *QS*, they must be parallel to each other.
   (c) It is a parallelogram, whatever the shape of the given quadrilateral.

B Components in two dimensions (p 116)

B1 B $\begin{bmatrix} 8 \\ -4 \end{bmatrix}$, C $\begin{bmatrix} 2 \\ -1 \end{bmatrix}$ and D $\begin{bmatrix} -16 \\ 8 \end{bmatrix}$

B2 $\begin{bmatrix} 14 \\ -6 \end{bmatrix}$

B3 $|\mathbf{p}| = 5$, $|\mathbf{q}| = 13$ and $|\mathbf{p} + \mathbf{q}| = \sqrt{68}$ or $2\sqrt{17}$.
   $|\mathbf{p}| + |\mathbf{q}| = 5 + 13 = 18 \neq \sqrt{68} = |\mathbf{p} + \mathbf{q}|$ so this verifies the fact that $|\mathbf{p}| + |\mathbf{q}| \neq |\mathbf{p} + \mathbf{q}|$ in general.

B4 (a) $\begin{bmatrix} 5 \\ 2 \end{bmatrix}$ (b) **x** + 3**y**

B5 −4**i** + 3**j**

B6 $\begin{bmatrix} 4 \\ 7 \end{bmatrix}$

B7 (a) 2**i** + 5**j**   (b) $\sqrt{29}$

B8 (a) 3**i** + 2**j**   (b) **i** + 8**j**   (c) 11**i**

B9 10

B10 Any multiple of 2**i** + 5**j**, e.g. 4**i** + 10**j**

B11 (a) 26
    (b) (i) 13   (ii) 2   (iii) $\frac{26}{3}$

B12 (a) 15   (b) 18**i** + 24**j**

B13 $\left|\frac{7}{25}\mathbf{i} + \frac{24}{25}\mathbf{j}\right| = \sqrt{\left(\frac{7}{25}\right)^2 + \left(\frac{24}{25}\right)^2}$
    $= \sqrt{\left(\frac{625}{625}\right)^2} = \sqrt{1} = 1$

B14 (a) 20   (b) $\frac{3}{5}\mathbf{i} + \frac{4}{5}\mathbf{j}$

Exercise B (p 119)

1 (a) $\begin{bmatrix} 1 \\ 8 \end{bmatrix}$ (b) $\begin{bmatrix} -1 \\ 12 \end{bmatrix}$ (c) $\begin{bmatrix} 0 \\ 0 \end{bmatrix}$

2 (a) $\sqrt{13}$ (b) 5
   (c) $\sqrt{2}$ (d) $\sqrt{130}$

3 (a) 3**i** + 2**j**   (b) $\sqrt{148}$ or $2\sqrt{37}$

4 $3\sqrt{26}$

5 **a** and **e**, **b** and **f**, **c** and **d**

6 ±2

7 $\begin{bmatrix} 9 \\ 12 \end{bmatrix}$ and $\begin{bmatrix} -9 \\ -12 \end{bmatrix}$

8 (a) $\left(\frac{5}{13}\right)^2 + \left(\frac{12}{13}\right)^2 = \frac{25}{169} + \frac{144}{169} = 1$ so the magnitude of the vector is $\sqrt{1}$ which is 1. Hence the vector is a unit vector.

(b) $\frac{3}{5}\mathbf{i} - \frac{4}{5}\mathbf{j}$

9 $k = 1, l = -4$

10 (a) (i) $\mathbf{a} + \mathbf{b}$　　**(ii)** $-2\mathbf{a}$　　**(iii)** $\frac{1}{2}\mathbf{b}$

(iv) $\frac{1}{2}\mathbf{b} - 2\mathbf{a}$　　**(v)** $\mathbf{b} - \mathbf{a}$

(b) (i) $\overrightarrow{PX}$ and $\overrightarrow{PR}$ are parallel and $\overrightarrow{PR} = \mathbf{a} + \mathbf{b}$.
Hence $\overrightarrow{PX} = k(\mathbf{a} + \mathbf{b})$ for some number k.
(ii) $\overrightarrow{XS}$ and $\overrightarrow{TS}$ are parallel and $\overrightarrow{TS} = \frac{1}{2}\mathbf{b} - 2\mathbf{a}$.
Hence $\overrightarrow{XS} = l(\frac{1}{2}\mathbf{b} - 2\mathbf{a})$ for some number l.
(iii) $k = \frac{3}{5}, l = \frac{4}{5}$

C Vectors in three dimensions (p 120)

C1 (a) (i) $\frac{1}{2}\mathbf{r}$　　**(ii)** $\mathbf{p} + \mathbf{q}$　　**(iii)** $\mathbf{p} + \mathbf{r}$
(iv) $\mathbf{q} - \mathbf{p}$　　**(v)** $\mathbf{p} + \mathbf{q} + \mathbf{r}$　　**(vi)** $-\mathbf{r} - \mathbf{p}$
(vii) $\mathbf{q} - \mathbf{p} + \frac{1}{2}\mathbf{r}$　　**(viii)** $\frac{1}{2}\mathbf{r} - \mathbf{q}$

(b) $\overrightarrow{AG} = \mathbf{q} + \frac{1}{2}\mathbf{r}$ and $\overrightarrow{DH} = \mathbf{q} + \mathbf{r}$.
$\mathbf{q} + \frac{1}{2}\mathbf{r}$ and $\mathbf{q} + \mathbf{r}$ are not multiples of each other and so the vectors are not parallel.

C2 (a) $(7, 0, 0)$　　**(b)** $(4, 8, 0)$
(c) $(7, 0, 6)$　　**(d)** $(4, 8, 6)$

C3 (a) 3　　**(b)** 8　　**(c)** 6

C4 (a) $\begin{bmatrix} 3 \\ 0 \\ 6 \end{bmatrix}$　**(b)** $\begin{bmatrix} 3 \\ 0 \\ 6 \end{bmatrix}$　**(c)** $\begin{bmatrix} 3 \\ 8 \\ 0 \end{bmatrix}$　**(d)** $\begin{bmatrix} -3 \\ 0 \\ 6 \end{bmatrix}$

(e) $\begin{bmatrix} 0 \\ 0 \\ -6 \end{bmatrix}$　**(f)** $\begin{bmatrix} 3 \\ 8 \\ 6 \end{bmatrix}$　**(g)** $\begin{bmatrix} -3 \\ 8 \\ 6 \end{bmatrix}$　**(h)** $\begin{bmatrix} -3 \\ 8 \\ -6 \end{bmatrix}$

C5 $\begin{bmatrix} 2 \\ -1 \\ 3 \end{bmatrix}, \begin{bmatrix} -6 \\ 3 \\ -9 \end{bmatrix}$ and $\begin{bmatrix} 4 \\ -2 \\ 6 \end{bmatrix}; \begin{bmatrix} 12 \\ -6 \\ 16 \end{bmatrix}$ and $\begin{bmatrix} 6 \\ -3 \\ 8 \end{bmatrix}$

C6 $2\mathbf{i} + 6\mathbf{j} - 5\mathbf{k}$

C7 $\begin{bmatrix} 3 \\ -2 \\ 8 \end{bmatrix}$

C8 $2\mathbf{i} + 4\mathbf{j} - 6\mathbf{k} = 2(\mathbf{i} + 2\mathbf{j} - 3\mathbf{k})$ so the vectors $2\mathbf{i} + 4\mathbf{j} - 6\mathbf{k}$ and $\mathbf{i} + 2\mathbf{j} - 3\mathbf{k}$ are parallel.

C9 (a) The dotted lines that represent the x- and y-components 1 and 3 are parallel to the x- and y-axes respectively. The x- and y-axes are at right angles to each other so the dotted lines must be too. Hence the shaded triangle is a right-angled triangle.

(b) $\sqrt{10}$　　**(c)** Yes　　**(d)** $\sqrt{14}$

C10 $\sqrt{178}$

C11 $\sqrt{42}$

C12 (a) 3　　**(b)** $\sqrt{35}$
(c) $\sqrt{45}$, or $3\sqrt{5}$

C13 (a) (i) $\begin{bmatrix} 4 \\ 0 \\ 0 \end{bmatrix}$　**(ii)** $\begin{bmatrix} 1 \\ 2 \\ 6 \end{bmatrix}$　**(iii)** $\begin{bmatrix} -1 \\ -2 \\ -6 \end{bmatrix}$

(b) $\overrightarrow{EB} = \overrightarrow{EA} + \overrightarrow{AB} = \begin{bmatrix} -1 \\ -2 \\ -6 \end{bmatrix} + \begin{bmatrix} 4 \\ 0 \\ 0 \end{bmatrix} = \begin{bmatrix} 3 \\ -2 \\ -6 \end{bmatrix}$

(c) (i) $4\mathbf{i} + 4\mathbf{j}$　　**(ii)** $3\mathbf{i} + 2\mathbf{j} - 6\mathbf{k}$
(iii) $\mathbf{i} - 2\mathbf{j} + 6\mathbf{k}$

(d) $\sqrt{41}$

C14 (a) (i) $\begin{bmatrix} 2 \\ 2 \\ 7 \end{bmatrix}$　**(ii)** $\begin{bmatrix} 14 \\ -2 \\ 2 \end{bmatrix}$　**(iii)** $\begin{bmatrix} 14 \\ -4 \\ 32 \end{bmatrix}$

(iv) $\begin{bmatrix} 5 \\ -6 \\ -2 \end{bmatrix}$　**(v)** $\begin{bmatrix} 4 \\ 6 \\ -16 \end{bmatrix}$

(b) $\sqrt{29}$

C15 (a) $|\mathbf{p}| = \sqrt{1^2 + 1^2 + 4^2} = \sqrt{18} = \sqrt{9} \times \sqrt{2} = 3\sqrt{2}$
(b) (i) $2\mathbf{i} + 5\mathbf{j}$　　**(ii)** $3\mathbf{i} - 4\mathbf{j} + 3\mathbf{k}$
(iii) $10\mathbf{j} + \mathbf{k}$
(c) $\sqrt{29}$

Exercise C (p 124)

1 (a) (i) $\begin{bmatrix} 0 \\ 15 \\ -2 \end{bmatrix}$　**(ii)** $\begin{bmatrix} 0 \\ -30 \\ 4 \end{bmatrix}$　**(iii)** $\begin{bmatrix} 0 \\ 0 \\ 0 \end{bmatrix}$

(b) 5

2 (a) 9　　**(b)** 27

3 (a) 3
(b) $\frac{1}{3}(2\mathbf{i} + 2\mathbf{j} - \mathbf{k})$ or $\frac{2}{3}\mathbf{i} + \frac{2}{3}\mathbf{j} - \frac{1}{3}\mathbf{k}$ or $\frac{1}{3}(-2\mathbf{i} - 2\mathbf{j} + \mathbf{k})$ or $-\frac{2}{3}\mathbf{i} - \frac{2}{3}\mathbf{j} + \frac{1}{3}\mathbf{k}$

4 (a) (i) $\begin{bmatrix} 3 \\ 0 \\ 0 \end{bmatrix}$ **(ii)** $\begin{bmatrix} 0 \\ 0 \\ 2 \end{bmatrix}$ **(iii)** $\begin{bmatrix} 0 \\ 3 \\ 0 \end{bmatrix}$

(b) $\overrightarrow{HB} = \overrightarrow{HE} + \overrightarrow{EA} + \overrightarrow{AB} = \begin{bmatrix} 0 \\ -3 \\ 0 \end{bmatrix} + \begin{bmatrix} 0 \\ 0 \\ -2 \end{bmatrix} + \begin{bmatrix} 3 \\ 0 \\ 0 \end{bmatrix}$

$= \begin{bmatrix} 3 \\ -3 \\ -2 \end{bmatrix}$

(c) (i) $2\mathbf{i} + 5\mathbf{j} + 2\mathbf{k}$ **(ii)** $5\mathbf{i} + 2\mathbf{j} - 2\mathbf{k}$

    (iii) $5\mathbf{i} + 5\mathbf{j} + 2\mathbf{k}$ **(iv)** $-5\mathbf{i} - 5\mathbf{j} + 2\mathbf{k}$

(d) $\sqrt{22}$

5 $\mathbf{u}$ and $\mathbf{y}$, $\mathbf{v}$ and $\mathbf{z}$, $\mathbf{w}$ and $\mathbf{x}$

6 $5\mathbf{i} + 8\mathbf{j}$

7 $x = 3, y = -6$

8 ± 4

9 $4\mathbf{i} - 6\mathbf{j} - 12\mathbf{k}$ or $-4\mathbf{i} + 6\mathbf{j} + 12\mathbf{k}$

10 You can use a proof by contradiction as follows.
Assume that it is possible to write $\mathbf{a}$ as a linear combination of $\mathbf{b}$ and $\mathbf{c}$. Then there must exist numbers t and s such that $\mathbf{a} = t\mathbf{b} + s\mathbf{c}$.
So $\mathbf{i} + \mathbf{j} - 2\mathbf{k} = t(5\mathbf{i} - \mathbf{j} + 3\mathbf{k}) + s(-2\mathbf{i} + \mathbf{j} + \mathbf{k})$
which gives the equations $5t - 2s = 1$, $-t + s = 1$ and $3t + s = -2$. The first two equations have the solution $t = 1, s = 2$. This gives $3t + s = 5$ which contradicts the fact that $3t + s = -2$. Hence the original assumption that it is possible to write $\mathbf{a}$ as a linear combination of $\mathbf{b}$ and $\mathbf{c}$ must be false.

D Position vectors in two and three dimensions (p 126)

D1 $\mathbf{q} - \mathbf{p}$

D2 (a) A sketch showing $X\,(1, 6)$ and $Y\,(2, -7)$

    (b) $(3, -1)$

    (c) $\mathbf{i} - 13\mathbf{j}$

D3 (a) $\frac{1}{4}(\mathbf{n} - \mathbf{m})$

    (b) The position vector of point P is $\overrightarrow{OP}$.
    $\overrightarrow{OP} = \overrightarrow{OM} + \overrightarrow{MP} = \mathbf{m} + \frac{1}{4}(\mathbf{n} - \mathbf{m})$
    $= \mathbf{m} + \frac{1}{4}\mathbf{n} - \frac{1}{4}\mathbf{m} = \frac{3}{4}\mathbf{m} + \frac{1}{4}\mathbf{n}$ as required.

D4 (a) $\begin{bmatrix} -2 \\ 8 \\ -6 \end{bmatrix}$ **(b)** $\sqrt{104}$ or $2\sqrt{26}$

Exercise D (p 129)

1 $\frac{4}{5}\mathbf{a} + \frac{1}{5}\mathbf{b}$

2 (a) $\mathbf{i} + 5\mathbf{j} + 7\mathbf{k}$ **(b)** $\sqrt{75}$ or $5\sqrt{3}$

3 (a) $O\begin{bmatrix} 0 \\ 0 \\ 0 \end{bmatrix}, A\begin{bmatrix} 4 \\ 0 \\ 0 \end{bmatrix}, B\begin{bmatrix} 4 \\ 4 \\ 0 \end{bmatrix}, C\begin{bmatrix} 0 \\ 4 \\ 0 \end{bmatrix}, D\begin{bmatrix} 0 \\ 0 \\ 4 \end{bmatrix}, E\begin{bmatrix} 4 \\ 0 \\ 4 \end{bmatrix},$

    $F\begin{bmatrix} 4 \\ 4 \\ 4 \end{bmatrix}, G\begin{bmatrix} 0 \\ 4 \\ 4 \end{bmatrix}$

    (b) $\overrightarrow{AG} = \begin{bmatrix} -4 \\ 4 \\ 4 \end{bmatrix}, \overrightarrow{BD} = \begin{bmatrix} -4 \\ -4 \\ 4 \end{bmatrix}$

    (c) $P\begin{bmatrix} 2 \\ 0 \\ 4 \end{bmatrix}, Q\begin{bmatrix} 4 \\ 0 \\ 2 \end{bmatrix}, R\begin{bmatrix} 4 \\ 2 \\ 4 \end{bmatrix}$

    (d) (i) $\overrightarrow{PQ} = \begin{bmatrix} 2 \\ 0 \\ -2 \end{bmatrix}, \overrightarrow{QR} = \begin{bmatrix} 0 \\ 2 \\ 2 \end{bmatrix}, \overrightarrow{RP} = \begin{bmatrix} -2 \\ -2 \\ 0 \end{bmatrix}$

    (ii) $\overrightarrow{PQ} + \overrightarrow{QR} + \overrightarrow{RP} = \begin{bmatrix} 0 \\ 0 \\ 0 \end{bmatrix}$ which is to be

    expected as PQ, QR and RP form the edges of a triangle.

4 $|\mathbf{a}| = \sqrt{5^2 + 5^2 + 4^2} = \sqrt{66}$

  $|\mathbf{b}| = \sqrt{4^2 + (-7)^2 + 1^2} = \sqrt{66}$

  $|\mathbf{c}| = \sqrt{7^2 + 4^2 + (-1)^2} = \sqrt{66}$

So points A, B and C are all the same distance from $(0, 0, 0)$ and hence on the surface of a sphere with centre $(0, 0, 0)$.

5 $\overrightarrow{PQ} = \mathbf{q} - \mathbf{p} = -\mathbf{i} + 2\mathbf{j} + 2\mathbf{k}$
  $\overrightarrow{RS} = \mathbf{s} - \mathbf{r} = 2\mathbf{i} - 4\mathbf{j} - 4\mathbf{k}$
  $2\mathbf{i} - 4\mathbf{j} - 4\mathbf{k} = -2(-\mathbf{i} + 2\mathbf{j} + 2\mathbf{k})$ so $\overrightarrow{RS} = -2\overrightarrow{PQ}$
  and hence the vectors are parallel.

6 (a) $\sqrt{133}$ **(b)** $\sqrt{114}$

7 One way is to argue that, since $\vec{AB} = 9\mathbf{i} + 3\mathbf{j} + 3\mathbf{k}$ and $\vec{BC} = -6\mathbf{i} - 2\mathbf{j} - 2\mathbf{k}$, then $\vec{BC} = -\frac{2}{3}\vec{AB}$. Hence BC is parallel to AB. So A, B and C are in a straight line.

8 $4, -2$

E The vector equation of a line (p 130)

E1 (a) When $t = 5$, $\begin{bmatrix} x \\ y \end{bmatrix} = \begin{bmatrix} 1 \\ 6 \end{bmatrix} + 5\begin{bmatrix} 2 \\ -1 \end{bmatrix}$

$= \begin{bmatrix} 1 \\ 6 \end{bmatrix} + \begin{bmatrix} 10 \\ -5 \end{bmatrix} = \begin{bmatrix} 11 \\ 1 \end{bmatrix}$ as required.

(b) (i) $\begin{bmatrix} 1 \\ 6 \end{bmatrix}$ **(ii)** $\begin{bmatrix} 3 \\ 5 \end{bmatrix}$ **(iii)** $\begin{bmatrix} -1 \\ 7 \end{bmatrix}$ **(iv)** $\begin{bmatrix} 7 \\ 3 \end{bmatrix}$

(c) (i) A diagram with these points plotted: $(-1, 7)$, $(1, 6)$, $(3, 5)$, $(7, 3)$ and $(11, 1)$.

(ii) Let P be the point $(1, 6)$. For any point Q with position vector given by the rule we have $\vec{PQ} = t\begin{bmatrix} 2 \\ -1 \end{bmatrix}$ for some t.
So all possible vectors $\vec{PQ}$ are parallel and hence all possible points Q are on the same straight line through P parallel to the vector $\begin{bmatrix} 2 \\ -1 \end{bmatrix}$.

(iii) $x + 2y = 13$

E2 (a) When $t = 3$, $\mathbf{r} = \mathbf{i} - \mathbf{j} + 3(\mathbf{i} + 2\mathbf{j}) = \mathbf{i} - \mathbf{j} + 3\mathbf{i} + 6\mathbf{j} = 4\mathbf{i} + 5\mathbf{j}$ as required.

(b) (i) $2\mathbf{i} + \mathbf{j}$ **(ii)** $\mathbf{i} - \mathbf{j}$
(iii) $-2\mathbf{i} - 7\mathbf{j}$ **(iv)** $6\mathbf{i} + 9\mathbf{j}$

(c) (i) A straight line through the points $(2, 1)$, $(1, -1)$, $(-2, -7)$, $(6, 9)$

(ii) $y = 2x - 3$

E3 (a) (i) $\mathbf{i} - \mathbf{j}$ **(ii)** $2\mathbf{i} + \mathbf{j}$ **(iii)** $5\mathbf{i} + 7\mathbf{j}$

(b) (i) A straight line through the points $(1, -1)$, $(2, 1)$ and $(5, 7)$

(ii) It is the same line.

(iii) The point $(2, 1)$ is on the line found in E1 and the vector $-\mathbf{i} - 2\mathbf{j}$ is parallel to the vector $\mathbf{i} + 2\mathbf{j}$. Hence the lines will be the same.

E4 Examples of equations are

(a) $\mathbf{r} = \begin{bmatrix} 0 \\ 4 \end{bmatrix} + t\begin{bmatrix} 1 \\ 1 \end{bmatrix}$ **(b)** $\mathbf{r} = \begin{bmatrix} 0 \\ 6 \end{bmatrix} + t\begin{bmatrix} 1 \\ -1 \end{bmatrix}$

(c) $\mathbf{r} = \begin{bmatrix} 0 \\ 1 \end{bmatrix} + t\begin{bmatrix} 1 \\ 2 \end{bmatrix}$

E5 (a)

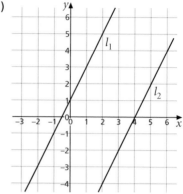

(b) (i) The lines are parallel.

(ii) The vectors $\begin{bmatrix} 1 \\ 2 \end{bmatrix}$ and $\begin{bmatrix} -2 \\ -4 \end{bmatrix}$ are parallel.

E6 L_2

E7 (a) Three points such as $(4, 3, 7)$, $(5, 3, 4)$, $(6, 3, 1)$, $(7, 3, -2)$, …

(b) $\lambda = 5$

(c) If the point does lie on the line then there must exist a number λ such that

$\begin{bmatrix} 7 \\ 3 \\ -1 \end{bmatrix} = \begin{bmatrix} 4 \\ 3 \\ 7 \end{bmatrix} + \lambda\begin{bmatrix} 1 \\ 0 \\ -3 \end{bmatrix}$. The value $\lambda = 3$ gives the

point with position vector $\begin{bmatrix} 7 \\ 3 \\ -2 \end{bmatrix}$ whose x- and

y-components match those of $\begin{bmatrix} 7 \\ 3 \\ -1 \end{bmatrix}$ but whose

z-component does not match. Hence no value

of λ exists so that $\begin{bmatrix} 7 \\ 3 \\ -1 \end{bmatrix} = \begin{bmatrix} 4 \\ 3 \\ 7 \end{bmatrix} + \lambda\begin{bmatrix} 1 \\ 0 \\ -3 \end{bmatrix}$ and so

the point with position vector $\begin{bmatrix} 7 \\ 3 \\ -1 \end{bmatrix}$ does not

lie on the line.

E8 (a) $7\mathbf{i} - 10\mathbf{j}$ (b) $\lambda = -1$

(c) If the point does lie on the line then there must exist a number λ such that
$3\mathbf{i} - 3\mathbf{j} - 2\mathbf{k} = \mathbf{i} + 2\mathbf{j} - 3\mathbf{k} + \lambda(2\mathbf{i} - 4\mathbf{j} + \mathbf{k})$.
The value $\lambda = 1$ gives the point with position vector $3\mathbf{i} - 2\mathbf{j} - 2\mathbf{k}$ whose x- and z-components match those of $3\mathbf{i} - 3\mathbf{j} - 2\mathbf{k}$ but whose y-component does not match.
Hence no value of λ exists such that
$3\mathbf{i} - 3\mathbf{j} - 2\mathbf{k} = \mathbf{i} + 2\mathbf{j} - 3\mathbf{k} + \lambda(2\mathbf{i} - 4\mathbf{j} + \mathbf{k})$ and so the point with position vector $3\mathbf{i} - 3\mathbf{j} - 2\mathbf{k}$ does not lie on the line.

Exercise E (p 135)

Each vector equation given in these answers is not unique but one of an infinite number of suitable equations.

1 (a) $\mathbf{r} = \begin{bmatrix} 0 \\ 5 \end{bmatrix} + \lambda \begin{bmatrix} 1 \\ 4 \end{bmatrix}$ (b) $\mathbf{r} = \lambda \begin{bmatrix} 1 \\ 0 \end{bmatrix}$

(c) $\mathbf{r} = 5\mathbf{i} - 2\mathbf{j} + \lambda(\mathbf{i} - 5\mathbf{j})$

2 (a) $\mathbf{r} = \begin{bmatrix} -1 \\ 3 \\ 5 \end{bmatrix} + \lambda \begin{bmatrix} -1 \\ 1 \\ -3 \end{bmatrix}$ (b) $\mathbf{r} = \begin{bmatrix} 2 \\ 1 \\ 0 \end{bmatrix} + \lambda \begin{bmatrix} 1 \\ 3 \\ 4 \end{bmatrix}$

(c) $\mathbf{r} = \lambda \begin{bmatrix} 0 \\ 0 \\ 1 \end{bmatrix}$

(d) $\mathbf{r} = 2\mathbf{i} - 4\mathbf{j} + 7\mathbf{k} + \lambda(-3\mathbf{i} + 6\mathbf{j} - 8\mathbf{k})$

3 When $\lambda = \frac{1}{2}$, $\mathbf{r} = \mathbf{i} - 3\mathbf{j} + 2\mathbf{k} + \frac{1}{2}(2\mathbf{i} + 4\mathbf{j} - 6\mathbf{k})$
$= \mathbf{i} - 3\mathbf{j} + 2\mathbf{k} + \mathbf{i} + 2\mathbf{j} - 3\mathbf{k} = 2\mathbf{i} - \mathbf{j} - \mathbf{k}$.
So the point with position vector $2\mathbf{i} - \mathbf{j} - \mathbf{k}$ lies on the line.

4 (a) $\mathbf{r} = 3\mathbf{i} + \mathbf{j} + 2\mathbf{k} + \lambda(8\mathbf{i} - 2\mathbf{j} - 4\mathbf{k})$

(b) For the equation above, when $\lambda = \frac{1}{2}$, $\mathbf{r} = 7\mathbf{i}$, which is the position vector of a point on the x-axis. So the line intersects the x-axis.

5 (a) $\mathbf{r} = -2\mathbf{i} + \mathbf{j} + \lambda(6\mathbf{i} - \mathbf{j} + 4\mathbf{k})$

(b) If the line did intersect the y-axis then there would exist numbers λ and a such that
$-2\mathbf{i} + \mathbf{j} + \lambda(6\mathbf{i} - \mathbf{j} + 4\mathbf{k}) = a\mathbf{j}$. Equating coefficients for the x- and z-components gives
$6\lambda - 2 = 0$ and $4\lambda = 0$. The first equation gives $\lambda = \frac{1}{3}$ but the second equation gives $\lambda = 0$.
Hence no such value for λ exists and the line does not intersect the y-axis.

6 (a) $\begin{bmatrix} 10 \\ 50 \\ 8 \end{bmatrix}$ (b) $16.3\,\text{km}$ (c) $1.3\,\text{km}$

(d) $0\,\text{km}$, which means the planes collide

7 $p = -17$, $q = 4$

8 $(5, 5)$, $(7, 1)$

9 $(2, -1, 2)$, $(1, -2, 2)$

F Intersecting lines (p 136)

F1 (a) For the first line we have $\begin{bmatrix} x \\ y \end{bmatrix} = \begin{bmatrix} 1 + \lambda \\ -3 + \lambda \end{bmatrix}$ and for the second we have $\begin{bmatrix} x \\ y \end{bmatrix} = \begin{bmatrix} 11 + \mu \\ 1 - 2\mu \end{bmatrix}$.
Equating x-components gives $1 + \lambda = 11 + \mu$ and equating y-components gives $-3 + \lambda = 1 - 2\mu$.

(b) $\lambda = 8$, $\mu = -2$

(c) $\begin{bmatrix} 9 \\ 5 \end{bmatrix}$

(d) Using the same parameter and equating the x- and y-components would only work if the parameter for each vector equation is the same at the point of intersection. Since the parameters are very likely to have different values at the point of intersection, we need to use different parameters to end up with the two different values.

F2 (a) For the first line we have $\mathbf{r} = 2\lambda\mathbf{i} + (2 + \lambda)\mathbf{j}$ and for the second we have
$\mathbf{r} = (3 + 4\mu)\mathbf{i} + (1 + 2\mu)\mathbf{j}$. Equating x-components gives $2\lambda = 3 + 4\mu$ and equating y-components gives $2 + \lambda = 1 + 2\mu$.

(b) There is no solution so the lines are parallel and do not intersect. This can be seen from the equations as the vector $\begin{bmatrix} 2 \\ 1 \end{bmatrix}$ is parallel to $\begin{bmatrix} 4 \\ 2 \end{bmatrix}$.

F3 $(1, 7)$

F4 There are an infinite number of solutions for λ and μ that are of the form $\lambda = -2 - 2\mu$. This means that the two lines are actually the same.

F5 (a) Parallel   (b) Intersecting   (c) Skew

    (d) Skew   (e) Parallel   (f) Skew

F6 For the first line we have $\mathbf{r} = \begin{bmatrix} 1+2\lambda \\ 2+3\lambda \\ 4-\lambda \end{bmatrix}$ and for the

second we have $\mathbf{r} = \begin{bmatrix} 3+4\mu \\ 6+5\mu \\ -2 \end{bmatrix}$.

Equating x-, y- and z-components gives the
equations $1 + 2\lambda = 3 + 4\mu$

$$2 + 3\lambda = 6 + 5\mu$$
$$4 - \lambda = -2$$

Solving simultaneously the first and third
equations gives $\lambda = 6$ and $\mu = 2\frac{1}{2}$ but this pair of
values gives $2 + 3\lambda = 20$ and $6 + 5\mu = 18\frac{1}{2}$ and so
does not satisfy the second equation. Hence the
lines do not intersect. Since they are not parallel,
they must be skew.

Exercise F (p 139)

1 If the two lines intersect then there must be values
of λ and μ such that $\begin{bmatrix} 5 \\ 4 \\ 3 \end{bmatrix} + \lambda \begin{bmatrix} 1 \\ 0 \\ -3 \end{bmatrix} = \begin{bmatrix} 7 \\ 10 \\ 9 \end{bmatrix} + \mu \begin{bmatrix} 0 \\ 2 \\ 4 \end{bmatrix}$

This gives the equations
$$5 + \lambda = 7$$
$$4 = 10 + 2\mu$$
$$3 - 3\lambda = 9 + 4\mu$$

The first and second equations give $\lambda = 2$ and
$\mu = -3$ and this pair of values gives $3 - 3\lambda = -3$
and $9 + 4\mu = -3$ too and so the values satisfy the
third equation. Hence the lines intersect.
The point of intersection is $(7, 4, -3)$.

2 If the two lines intersect then there must be values
of t and s such that $\begin{bmatrix} 2 \\ 0 \\ -1 \end{bmatrix} + t \begin{bmatrix} 0 \\ 2 \\ 3 \end{bmatrix} = \begin{bmatrix} -3 \\ 0 \\ 2 \end{bmatrix} + s \begin{bmatrix} 5 \\ 4 \\ -1 \end{bmatrix}$.

This gives the equations
$$2 = -3 + 5s$$
$$2t = 4s$$
$$-1 + 3t = 2 - s$$

The first and second equations give $s = 1$ and
$t = 2$ but this pair of values gives $-1 + 3t = 5$ and
$2 - s = 1$ and so the values do not satisfy the third
equation. Hence the lines do not intersect. Since
their direction vectors are not scalar multiples of
each other, the lines are not parallel and so they
must be skew.

3 If the two lines intersect then there must be values
of λ and μ such that $4\mathbf{i} + 3\mathbf{j} + \lambda(-\mathbf{i} - \mathbf{j} + \mathbf{k})$
$= 4\mathbf{i} - 3\mathbf{j} + \mu(-\mathbf{i} + \mathbf{j} + \mathbf{k})$. This gives the equations
$$4 - \lambda = 4 - \mu$$
$$3 - \lambda = -3 + \mu$$
$$\lambda = \mu$$

The first and second equations give $\lambda = 3$ and
$\mu = 3$ and clearly this pair of values satisfies the
third equation. Hence the lines intersect.
The point of intersection is $(1, 0, 3)$.

4 (a) Intersect at $(5, 3, 11)$

   (b) Parallel

   (c) Skew

   (d) Intersect at $(6, 3, -2)$

   (e) Skew

5 (a) 2   (b) $(2, 0, 2)$

6 $a = 3$, $(\frac{1}{2}, 3, -\frac{3}{2})$

7 (a) 1   (b) $(5, 4, 4)$

G Angles and the scalar product (p 140)

G1 (a) 90°   (b) 45°   (c) 135°

G2 (a) Vectors $\mathbf{a}$ and $\mathbf{b}$ can be drawn in any position
      and could be shown placed 'head to head'.

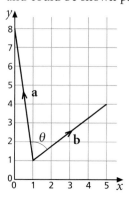

    (b) $|\mathbf{a}| = \sqrt{50}$ or $5\sqrt{2}$, $|\mathbf{b}| = 5$

(c) $\sqrt{41}$

(d) 61.3° (to 1 d.p.)

G3 (a) $|\mathbf{p}| = 3$, $|\mathbf{q}| = \sqrt{41}$

(b) $\sqrt{14}$

(c) 20.4° (to 1 d.p.)

G4 (a) −5

(b) $|\mathbf{a}| = \sqrt{14}$, $|\mathbf{b}| = \sqrt{50}$ or $5\sqrt{2}$

(c) $\cos\theta = -\dfrac{1}{2\sqrt{7}}$, $\theta = 100.9°$ (to 1 d.p.)

G5 $\mathbf{a}.\mathbf{b} = (2\times1) + (1\times-1) + (4\times1) = 2 - 1 + 4 = 5$

G6 $\overrightarrow{OA}.\overrightarrow{OB} = |\overrightarrow{OA}||\overrightarrow{OB}|\cos\angle AOB$
$= 5\times5\times\cos35 = 20.48$ (to 2 d.p.)

G7 (a) 0   **(b)** $\cos\theta = 0$

(c) They are at right angles to each other, that is they are perpendicular.

G8 The scalar product of the two vectors is
$(2\times4) + (-3\times1) + (1\times-5) = 8 - 3 - 5 = 0$ so the two vectors are perpendicular.

Exercise G (p 142)

1 (a) $\mathbf{a}.\mathbf{b} = 0$, $\mathbf{b}.\mathbf{c} = 0$, $\mathbf{a}.\mathbf{c} = 20$

(b) $\mathbf{a}$ and $\mathbf{b}$, $\mathbf{b}$ and $\mathbf{c}$

2 To 1 d.p., unless exact, the angles are

(a) 73.9°   **(b)** 132.8°   **(c)** 90°

3 79.3° (to 1 d.p.)

4 $\angle QPR = 90°$, $\angle PQR = 45°$, $\angle QRP = 45°$

5 Using the formula for the scalar product gives
$\mathbf{a}.\mathbf{a} = (a_1 \times a_1) + (a_2 \times a_2) + (a_3 \times a_3)$
$= a_1^2 + a_2^2 + a_3^2$
Using the formula for the length of a vector gives
$|\mathbf{a}|^2 = a_1^2 + a_2^2 + a_3^2$ too so $\mathbf{a}.\mathbf{a} = |\mathbf{a}|^2$.
Another way to show this is to note that the angle between a vector and itself is 0 so
$\mathbf{a}.\mathbf{a} = |\mathbf{a}||\mathbf{a}|\cos0 = |\mathbf{a}||\mathbf{a}| \times 1 = |\mathbf{a}|^2$.

6 Using the formula $\mathbf{a}.\mathbf{b} = |\mathbf{a}||\mathbf{b}|\cos\theta$ with the values $\mathbf{a}.\mathbf{b} = 12$, $|\mathbf{a}| = 3$ and $|\mathbf{b}| = 4$ gives
$12 = 3\times4\times\cos\theta$ which implies that $\theta = 0°$.
Hence $\mathbf{a}$ and $\mathbf{b}$ are parallel.

7 Labelling the points O (0, 0, 0), A (4, −2, 5), B (3, 6, 0) and C (7, 4, 5) we obtain these vectors.

$$\overrightarrow{OA} = \begin{bmatrix} 4 \\ -2 \\ 5 \end{bmatrix} \quad \overrightarrow{OB} = \begin{bmatrix} 3 \\ 6 \\ 0 \end{bmatrix} \quad \overrightarrow{OC} = \begin{bmatrix} 7 \\ 4 \\ 5 \end{bmatrix}$$

$$\overrightarrow{AB} = \begin{bmatrix} -1 \\ 8 \\ -5 \end{bmatrix} \quad \overrightarrow{AC} = \begin{bmatrix} 3 \\ 6 \\ 0 \end{bmatrix} \quad \overrightarrow{BC} = \begin{bmatrix} 4 \\ -2 \\ 5 \end{bmatrix}$$

We can see that $\overrightarrow{OA} = \overrightarrow{BC}$ and $\overrightarrow{OB} = \overrightarrow{AC}$ so we know that $OBCA$ is a parallelogram.
$|\overrightarrow{OA}| = \sqrt{4^2 + (-2)^2 + 5^2} = \sqrt{45}$ and
$|\overrightarrow{OB}| = \sqrt{3^2 + 6^2 + 0^2} = \sqrt{45}$ so $OBCA$ is a rhombus.

Finally, to show that $OBCA$ is a square we need to show that one of its angles is a right angle (which implies here that all the angles are right angles).
$\overrightarrow{OB}.\overrightarrow{OA} = (3\times4) + (6\times-2) + (0\times5) = 0$ so $\angle BOA = 90°$ and so the vertices form a square.

There are several valid alternative ways of answering this question.

H The angle between two straight lines (p 143)

H1 (a) No

(b) You could move the line through AC vertically upwards till it meets HB and measure the angle between them.

H2 90°

H3 (a) 90°   **(b)** 45°   **(c)** 90°   **(d)** 60°

H4 The angle between the lines is the angle between

the direction vectors $\begin{bmatrix} 3 \\ 5 \\ -1 \end{bmatrix}$ and $\begin{bmatrix} 1 \\ 2 \\ 13 \end{bmatrix}$. So to show

that the lines are perpendicular you can show that

$\begin{bmatrix} 3 \\ 5 \\ -1 \end{bmatrix}.\begin{bmatrix} 1 \\ 2 \\ 13 \end{bmatrix} = 0$

Exercise H (p 145)

1 (a) 60°   **(b)** 60°   **(c)** 45°   **(d)** 90°

2 30.6° (to 1 d.p.)

3 The scalar product of the direction vectors is
$(3\mathbf{i} - 6\mathbf{j}).(2\mathbf{i} + \mathbf{j} - 3\mathbf{k}) = (3 \times 2) + (-6 \times 1) + (0 \times -3)$
$= 6 - 6 + 0 = 0$ so the two lines are perpendicular.

4 $81.6°$ (to 1 d.p.)

5 $88.2°$ (to 1 d.p.)

I Shortest distance (p 146)

I1 θ is $90°$.

Exercise I (p 147)

1 $(5.7, 1.9)$

2 $(-2, 5)$

3 $\sqrt{40}$, or $2\sqrt{10}$

4 $(4, \frac{1}{2}, -\frac{5}{2})$

5 $\sqrt{126}$, or $3\sqrt{14}$

6 $(1, 1, 3)$

7 **(a)** One possible equation is
$\mathbf{r} = 2\mathbf{i} - 3\mathbf{k} + \lambda(\mathbf{i} + \mathbf{j} + 2\mathbf{k})$.

 (b) $\sqrt{54}$ or $3\sqrt{6}$

 (c) The shortest distance is $\sqrt{3}$ and the area is
$\dfrac{\sqrt{162}}{2}$ or $\dfrac{3\sqrt{18}}{2}$ or $\dfrac{9\sqrt{2}}{2}$ or $\dfrac{9}{\sqrt{2}}$.

Mixed questions (p 150)

Each vector equation given in these answers is not unique but one of an infinite number of suitable equations.

1 **(a)** If the two lines intersect then there must be values of λ and μ such that
$2\mathbf{i} + \mathbf{j} + \lambda(\mathbf{i} + \mathbf{j} - 2\mathbf{k}) = 5\mathbf{i} + 2\mathbf{j} - \mathbf{k} + \mu(2\mathbf{i} + \mathbf{k})$.
This gives the equations
$$2 + \lambda = 5 + 2\mu$$
$$1 + \lambda = 2$$
$$-2\lambda = -1 + \mu$$

 The first and second equations give $\lambda = 1$ and $\mu = -1$ and this pair of values gives $-2\lambda = -2$ and $-1 + \mu = -2$ too and so the values satisfy the third equation. Hence the lines intersect. The point of intersection is $B\,(3, 2, -2)$.

(b) The scalar product of the direction vectors is
$\begin{bmatrix} 1 \\ 1 \\ -2 \end{bmatrix}.\begin{bmatrix} 2 \\ 0 \\ 1 \end{bmatrix} = (1 \times 2) + (1 \times 0) + (-2 \times 1)$
$= 2 + 0 - 2 = 0$ so the lines are perpendicular.

(c) When $\lambda = -1$, $\mathbf{r} = \begin{bmatrix} 2 \\ 1 \\ 0 \end{bmatrix} + -1\begin{bmatrix} 1 \\ 1 \\ -2 \end{bmatrix} = \begin{bmatrix} 2 \\ 1 \\ 0 \end{bmatrix} + \begin{bmatrix} -1 \\ -1 \\ 2 \end{bmatrix}$
$= \begin{bmatrix} 1 \\ 0 \\ 2 \end{bmatrix}$. So point $A\,(1, 0, 2)$ lies on the line.

(d) $p = 2$, $q = -3$

(e) $\sqrt{30}$

2 **(a)** $\mathbf{r} = \begin{bmatrix} 1 \\ 1 \\ -2 \end{bmatrix} + \lambda\begin{bmatrix} 1 \\ -4 \\ 2 \end{bmatrix}$

 (b) $\mathbf{r} = \mu\begin{bmatrix} 7 \\ -3 \\ 8 \end{bmatrix}$

 (c) A direction vector for l_1 is $\begin{bmatrix} 1 \\ -4 \\ 2 \end{bmatrix}$ and a
direction vector for l_2 is $\begin{bmatrix} 7 \\ -3 \\ 8 \end{bmatrix}$. They are not
scalar multiples of each other, so the lines are not parallel.

 If the two lines intersect then there must be values of λ and μ such that
$\begin{bmatrix} 1 \\ 1 \\ -2 \end{bmatrix} + \lambda\begin{bmatrix} 1 \\ -4 \\ 2 \end{bmatrix} = \mu\begin{bmatrix} 7 \\ -3 \\ 8 \end{bmatrix}$.

 This gives the equations
$$1 + \lambda = 7\mu$$
$$1 - 4\lambda = -3\mu$$
$$-2 + 2\lambda = 8\mu$$

 The first and second equations give $\lambda = \frac{2}{5}$ and $\mu = \frac{1}{5}$ but this pair of values gives $-2 + 2\lambda = -\frac{6}{5}$ and $8\mu = \frac{8}{5}$ and so the values do not satisfy the third equation. Hence the lines do not intersect.

 Hence the lines are skew.

 (d) $\sqrt{68}$, or $2\sqrt{17}$

3 (a) $\mathbf{r} = 5\mathbf{i} + 4\mathbf{k} + \lambda(3\mathbf{i} + \mathbf{j} - \mathbf{k})$

(b) $p = 8$, $q = 3$

(c) $39.5°$ (to 1 d.p.)

(d) $2\mathbf{i} - \mathbf{j} + 5\mathbf{k}$

4 (a) $\mathbf{r} = 2\mathbf{i} + \mathbf{j} + \mathbf{k} + \lambda(\mathbf{i} - \mathbf{k})$

(b) If lines intersect then there must be values of λ and μ such that $2\mathbf{i} + \mathbf{j} + \mathbf{k} + \lambda(\mathbf{i} - \mathbf{k})$
$= 3\mathbf{j} - \mathbf{k} + \mu(-2\mathbf{i} + \mathbf{j})$.

This gives the equations
$$2 + \lambda = -2\mu$$
$$1 = 3 + \mu$$
$$1 - \lambda = -1$$

The first and second equations give $\lambda = 2$ and $\mu = -2$ and the value for λ gives $1 - \lambda = -1$ and so the values satisfy the third equation. Hence the lines intersect.
The point of intersection is $(4, 1, -1)$.

(c) When $\mu = 4$, $\mathbf{r} = \begin{bmatrix} 0 \\ 3 \\ -1 \end{bmatrix} + 4\begin{bmatrix} -2 \\ 1 \\ 0 \end{bmatrix} = \begin{bmatrix} 0 \\ 3 \\ -1 \end{bmatrix} + \begin{bmatrix} -8 \\ 4 \\ 0 \end{bmatrix}$

$= \begin{bmatrix} -8 \\ 7 \\ -1 \end{bmatrix}$. So point C $(-8, 7, -1)$ lies on the line.

(d) $(-2, 1, 5)$

5 $88.5°$ (to 1 d.p.)

Test yourself (p 151)

1 (a) $\overrightarrow{AB} = (5\mathbf{i} + 7\mathbf{j} + 4\mathbf{k}) - (2\mathbf{i} + \mathbf{j} + \mathbf{k})$
$= 3\mathbf{i} + 6\mathbf{j} + 3\mathbf{k}$
$\overrightarrow{AC} = (\mathbf{i} - \mathbf{j}) - (2\mathbf{i} + \mathbf{j} + \mathbf{k})$
$= -\mathbf{i} - 2\mathbf{j} - \mathbf{k}$
$= -\tfrac{1}{3}(3\mathbf{i} + 6\mathbf{j} + 3\mathbf{k})$
$= -\tfrac{1}{3}\overrightarrow{AB}$

Hence A, B and C lie in a straight line.

(b) $\dfrac{1}{\sqrt{84}}$ or $\dfrac{1}{2\sqrt{21}}$

(c) $\overrightarrow{AE} = (-3\mathbf{j} - \mathbf{k}) - (2\mathbf{i} + \mathbf{j} + \mathbf{k})$
$= -2\mathbf{i} - 4\mathbf{j} - 2\mathbf{k}$
$= -\tfrac{2}{3}\overrightarrow{AB}$
So E is on l.
$\overrightarrow{OE}.\overrightarrow{OD} = (-3\mathbf{j} - \mathbf{k}).(2\mathbf{i} + \mathbf{j} - 3\mathbf{k})$
$= (0 \times 2) + (-3 \times 1) + (-1 \times -3)$
$= -3 + 3 = 0$
So OE and OD are perpendicular.

2 (a) $\mathbf{r} = 3\mathbf{i} + \mathbf{j} - 2\mathbf{k} + \lambda(8\mathbf{i} + \mathbf{j} - 3\mathbf{k})$

(b) Let P be the point on the line l that is closest to Q $(6, 5, -5)$ and let O be the origin $(0, 0, 0)$. As P is on the line, there must be some value of λ for which
$\overrightarrow{OP} = 3\mathbf{i} + \mathbf{j} - 2\mathbf{k} + \lambda(8\mathbf{i} + \mathbf{j} - 3\mathbf{k})$
$= (3 + 8\lambda)\mathbf{i} + (1 + \lambda)\mathbf{j} + (-2 - 3\lambda)\mathbf{k}$
$\overrightarrow{QP} = \overrightarrow{OP} - \overrightarrow{OQ} = ((3 + 8\lambda)\mathbf{i} + (1 + \lambda)\mathbf{j} + (-2 - 3\lambda)\mathbf{k}) - (6\mathbf{i} + 5\mathbf{j} - 5\mathbf{k})$
$= (-3 + 8\lambda)\mathbf{i} + (-4 + \lambda)\mathbf{j} + (3 - 3\lambda)\mathbf{k}$

The direction of l is given by the vector $8\mathbf{i} + \mathbf{j} - 3\mathbf{k}$ and we know that $\overrightarrow{QP}$ must be perpendicular to this vector. So their scalar product must be 0.

Hence
$((-3 + 8\lambda)\mathbf{i} + (-4 + \lambda)\mathbf{j} + (3 - 3\lambda)\mathbf{k}).(8\mathbf{i} + \mathbf{j} - 3\mathbf{k})$
$= 0$
$\Rightarrow 8(-3 + 8\lambda) + (-4 + \lambda) - 3(3 - 3\lambda) = 0$
$\Rightarrow -24 + 64\lambda - 4 + \lambda - 9 + 9\lambda = 0$
$\Rightarrow 74\lambda - 37 = 0$
$\Rightarrow \lambda = \tfrac{1}{2}$
So $\overrightarrow{QP} = (-3 + 8 \times \tfrac{1}{2})\mathbf{i} + (-4 + \tfrac{1}{2})\mathbf{j} + (3 - 3 \times \tfrac{1}{2})\mathbf{k}$
$= \mathbf{i} - \tfrac{7}{2}\mathbf{j} + \tfrac{3}{2}\mathbf{k}$
The shortest distance is
$|\overrightarrow{QP}| = \sqrt{1^2 + \left(-\dfrac{7}{2}\right)^2 + \left(\dfrac{3}{2}\right)^2}$
$= \sqrt{1 + \dfrac{49}{4} + \dfrac{9}{4}}$
$= \sqrt{\dfrac{62}{4}} = \dfrac{\sqrt{62}}{2} = \tfrac{1}{2}\sqrt{62}$ as required.

3 (a) (i) $\frac{-4}{21}$

(ii) $\cos \angle PQR$ is negative but since the angle PQR is part of a triangle then it must be less than $180°$ and so $\sin \angle PQR$ must be positive. Hence

$$\sin \angle PQR = \sqrt{1 - (\cos \angle PQR)^2}$$

$$= \sqrt{1 - \left(\frac{-4}{21}\right)^2} = \sqrt{1 - \frac{16}{441}} = \sqrt{\frac{425}{441}} = \frac{5\sqrt{17}}{21}$$

as required.

(b) $\dfrac{5\sqrt{17}}{2}$

4 (a) If the two lines intersect then there must be values of t and s such that

$$5\mathbf{i} + \mathbf{j} - \mathbf{k} + t(2\mathbf{i} + \mathbf{j} + 5\mathbf{k})$$
$$= 13\mathbf{i} - 6\mathbf{j} + 2\mathbf{k} + s(-3\mathbf{i} + 4\mathbf{j} + \mathbf{k}).$$

This gives the equations

$$5 + 2t = 13 - 3s$$
$$1 + t = -6 + 4s$$
$$-1 + 5t = 2 + s$$

The first and second equations give $t = 1$ and $s = 2$ and this pair of values gives $-1 + 5t = 4$ and $2 + s = 4$ too and so the values satisfy the third equation. Hence the lines intersect. The point of intersection has coordinates $(7, 2, 4)$.

(b) $(9, 3, 9)$

5 (a) $\alpha = -3$

(b) $-15\mathbf{i} - 7\mathbf{j} + 19\mathbf{k}$

(c) The scalar product of the direction vectors is

$$(-8\mathbf{i} - 3\mathbf{j} + 5\mathbf{k}).(\mathbf{i} - 4\mathbf{j} + 9\mathbf{k})$$
$$= -8 + 12 + 45 = 49$$
$$|-8\mathbf{i} - 3\mathbf{j} + 5\mathbf{k}| = \sqrt{98} = 7\sqrt{2}$$
$$|\mathbf{i} - 4\mathbf{j} + 9\mathbf{k}| = \sqrt{98} = 7\sqrt{2}$$

So, if θ is the angle between the direction vectors, then $\cos \theta = \dfrac{49}{7\sqrt{2} \times 7\sqrt{2}} = \frac{1}{2}$ so $\theta = 60°$

Hence the acute angle between the line is $60°$.

(d) $\overrightarrow{OB} = \mathbf{i} - \mathbf{j} + 9\mathbf{k}$ and $\overrightarrow{OC} = -17\mathbf{i} + \mathbf{j} + \mathbf{k}$ or

$$\overrightarrow{OB} = -31\mathbf{i} - 13\mathbf{j} + 29\mathbf{k} \text{ and}$$
$$\overrightarrow{OC} = -13\mathbf{i} - 15\mathbf{j} + 37\mathbf{k}$$

Index